Biosphere Reserves and Sustainable Development Goals 1

Book published within the framework of the Erasmus+ "EduBioMed"

This project has been funded with support from the European Union. This publication reflects only the author's view and the Agency and the Commission are not responsible for any use that may be made of the information it contains.

Project number: 598924-EPP-1-2018-1-ES-EPPKA2-CBHE-JP

Co-funded by the Erasmus+ Programme of the European Union

EduBioMed

Territory Development Set

coordinated by
Angela Barthes

Volume 3

Biosphere Reserves and Sustainable Development Goals 1

Scientific and Practical Educational Issues in the Mediterranean

Edited by
Angela Barthes
Catherine Cibien
Bruno Romagny

iSTE

WILEY

First published 2023 in Great Britain and the United States by ISTE Ltd and John Wiley & Sons, Inc.

ISTE Ltd
27-37 St George's Road
London SW19 4EU
UK

www.iste.co.uk

John Wiley & Sons, Inc.
111 River Street
Hoboken, NJ 07030
USA

www.wiley.com

Library of Congress Control Number: 2023943622

British Library Cataloguing-in-Publication Data
A CIP record for this book is available from the British Library
ISBN 978-1-78630-780-4

Contents

Presentation of the Authors of the Two Volumes

Hannah Abou Fakher holds a bachelor's degree in Agricultural Studies and a master's degree in Environmental Engineering and Natural Resources. She has worked on various projects as an intern at the Nature Conservation Center at the American University of Beirut (AUB). Hannah is currently working with the International Federation of the Red Cross and Red Crescent Societies in the area of climate and resilience.

Mohammed Aderghal is a geographer; a higher education professor at the Faculty of Letters and Human Sciences of Mohammed V University in Rabat; Director of the Laboratory of Tourism Engineering, Heritage and Sustainable Development of the Territories (LITOPAD); President of the Association of Moroccan Geographers (ANAGEM); and founding member of the international joint laboratory (LMI) Mediterranean Terroirs: Heritage, Mobility, Change and Social Innovation (MédiTer). He works on the agro-pastoral and rural changes at the origin of territorial recompositions and the change in the relationship between society and natural resources in the mountains and in the pre-Saharan and Saharan regions.

Sonia Aderghal holds a PhD in Geography from the Faculty of Letters and Human Sciences (Mohammed V University in Rabat). A researcher and associate at the LITOPAD and LMI MédiTer, her work focuses on the territorial dynamics of Mediterranean coasts and their hinterlands. She focuses on themes relating to tourism, urbanization, new ruralities and the governance of territories.

Abdelaziz Afker is a water and forest engineer (1988, Salé, Morocco) and holds a master's degree in Human Resources and Organization Management (Nancy 2009). The focal point for the Arganeraie biosphere reserve at the Regional

Directorate of Water and Forests in Agadir, he has held several positions of responsibility related to management, studies and partnership for the benefit of natural resources.

Abdullah Aït L'Houssain is a PhD student in Geography at the Laboratory of Studies on Resources, Mobility and Attractiveness (LERMA) at Cadi-Ayyad University in Marrakech. The title of his thesis is "Climate Change, Migration and Sustainable Development". His areas of research are climate migration, social movement and territorial feminism.

Lahoucine Amzil is based at Mohammed V University, Rabat. He holds a dual education in Social Sciences (Mohammed V University, Faculty of Arts and Letters, 1998–2002) and in Ecological Sciences (Albert Ludwig-Freiburg University, Department of Geobotany and Forestry, 2002–2005). Lahoucine Amzil's doctoral research focuses on recent transformations of the traditional socio-economic system of the western High Atlas. Amzil's current research interests include sustainable approaches, migration processes and rural development.

Carla Aramony is an architect and assistant professor at AUB.

Nina Asloum is an associate research professor (HDR) in Education and Training Sciences at the National School of Agricultural Education in Toulouse. She is responsible for the training of teachers in Agricultural Education for Spatial Planning. Her research focuses on curricula for producing differently and on the socio-historical evolution of training curricula for agricultural education within the Toulouse UMR Education, Training, Work, Knowledge (EFTS) and for the theme Changes in Education and Training: Engagement, Interactions, Emancipation.

Hicham Attouch is a professor at Mohammed V University, Rabat, in the Faculty of Legal, Economic and Social Sciences.

Lahcen Azougarh is a researcher at the Environment, Development and Spatial Management Laboratory at Ibn Tofail University, Kenitra, Morocco.

Didier Babin is president of the French committee of UNESCO's Man and the Biosphere (MAB) Programme and was president of the International Coordinating Council from 2016 to 2018. He is a member of the French National Biodiversity Committee. He is currently responsible for the Post 2020 Biodiversity Framework EU Support project at Expertise France. A doctor in Geography, researcher at CIRAD and associate at the University of Quebec in Montreal, Didier Babin was part of the CBD secretariat for five years as head of the Biodiversity for Development and Poverty Eradication program and was a member of the United Nations technical team for the preparation of the Sustainable Development Goals.

He was also involved in the emergence of the Intergovernmental Science-Policy Platform on Biodiversity and Ecosystem Services (IPBES) as executive secretary of the IMoSEB process.

Joelle Barakat is a conservation manager at the Jabal Moussa biosphere reserve, Lebanon.

Angela Barthes is a professor at the University of Aix-Marseille and specializes in the field of environmental education and the development of rural territories. She leads the Education, Territories, Development, Society, Health team from the ADEF laboratory.

Laurent Bedoussac is an associate research professor (HDR) in Agronomy at the National School of Agricultural Education in Toulouse. He leads the training of teachers in horticultural production and viticulture. He conducts research in the UMR AGIR at the National Research Institute for Agriculture, Food and the Environment on the analysis of the performance and functioning of mixtures of species for the design of agroecological cropping systems.

Sylviane Blanc-Maximin is a doctor with an HDR in Education Sciences specializing in the links between territory and education. She is a teacher trainer at the University of Aix-Marseille and a member of the ADEF laboratory.

Martí Boada is a professor at the Autonomous University of Barcelona (UAB), a member of the Institute of Environmental Sciences and Technologies (ICTA), a doctor of environmental sciences, the author of over 100 popular science books and a pioneer in ecology.

Antonio Bontempi is the project manager for EduBioMed and a doctoral student at the department of Geography at UAB. He holds a joint European MSc in Environmental Studies (2017) and was awarded another MSc in Building Engineering from the University of Bergamo (Italy 2013). He has been a visiting professor at several universities: the University of Columbia in New York (2016), the Technical University of Hamburg (2016) and the Instituto Superior Técnico de Lisboa (2012). Antonio's interests include protected areas, socio-environmental studies, ecological economics, political ecology, planning and territorial development.

Meriem Bouamrane is an environmental economist and Head of the Biodiversity and Ecology section within the Division of Ecological and Earth Sciences in the Man and the Biosphere (MAB) Programme. She is responsible for research and training programs on the access to and use of biodiversity, participatory approaches and consultation, using biosphere reserves as research and demonstration sites. She is the biodiversity focal point within UNESCO, as well as for IPBES and Future Earth.

Éliane Bou Dagher is a member of the Faculty of Sciences at Saint-Joseph University in Beirut (USJ).

Magda Bou Dagher Kharrat is a professor at the Faculty of Sciences at USJ in Beirut. Magda is the author of numerous scientific publications on Lebanon's biodiversity and the countless threats it faces. She is also the founder and president of the NGO Jouzour Loubnan.

Saïd Boujrouf is a professor of Geography, director of the Laboratory for Studies on Resources, Mobility and Attractiveness (LERMA) and co-director of the LMI MédiTer at Cadi-Ayyad University, Marrakesh. He studies land use planning and territorial development in Morocco.

Soukaina Bouziani, after finishing her baccalaureate studies in Physical Sciences, turned to Economics and Management Studies, successfully obtaining her license. Following short professional experiences in administration during and after her studies, Soukaina chose to complete her academic career with a master's degree in Governance, Territorial Planning, Local and Regional Development and Resource Management. In her final project, she addressed the themes of sustainable economy and the promotion of biodiversity in the Moroccan part of the Intercontinental Biosphere Reserve of the Mediterranean (IBRM).

Abdelkarim Brahmi is a lecturer at the Higher Institute of Applied Studies in Humanities at the University of Gafsa. He holds a doctorate in Heritage Sciences from FSHST, University of Tunis. He is a member of the Laboratory of Maghreb Studies at the same faculty.

Driss Chahhou is a professor at the Faculty of Sciences of Mohammed V University, Rabat.

Abdelkader Chahlaoui is a professor at the Faculty of Sciences of Moulay Ismail University in Meknes, head of the Valorization of Natural Resources Management research team and author and co-author of 100 scientific articles in peer-reviewed journals. He is also an expert for national and international bodies and initiator of scientific research projects.

Véronique Chalando is a doctoral student in the field of education for sustainable development; her area of research is agroecological knowledge. A certified professor of agricultural education in ecology and biology, she now works in agronomy at the University Institute of Technology of the University of Aix-Marseille.

Mikaël Chambru is a lecturer in Information and Communication Sciences at Grenoble Alpes University and researcher within the laboratory for excellence (Labex) Innovation and Territorial Transitions in the Mountains (ITTEM) and the Research Group on Communication Issues (Gresec). His work focuses on the methods for publicizing science in mountain territories, looking at mediation, communication and popularization of science, as well as the controversies and public challenges regarding Alpine societal issues in the context of a general demand for an ecological transition.

Catherine Cibien is the director of the French committee of the MAB Programme of UNESCO. She supervises the national network of biosphere reserves, including a set of working groups and projects relating to the implementation of the Sustainable Development Goals (biodiversity, education, communication, commitment of actors, etc.); acts as liaison with the network world; and takes part in the bodies of the Programme at UNESCO. She co-supervises the MAB masters at the University of Toulouse and chairs the Scientific Council of the Cévennes National Park biosphere reserve.

Cécilia Claeys is an associate research professor (HDR) of sociology at the University of Perpignan and a member of the Population–Environment–Development Laboratory. A sociologist, she conducts interdisciplinary research combining human sciences and life sciences. Her preferred fields of study are in mainland and overseas France. Her work focuses on the implementation of public environmental and risk prevention policies and their (non-)acceptance by populations and territorial actors.

Tarik Dahou is a research director at UMR 208 PALOC, IRD, National Museum of Natural History, Paris.

Mchich Derak is a forestry engineer at DREFLCD, Rif, Morocco.

Pierre Doumet is co-founder and president of the Association for the Protection of Jabal Moussa and manager of the Jabal Moussa biosphere reserve, Lebanon.

El Habib El Azzouzi is a professor at the Scientific Institute of Mohammed V University, Rabat, specialist in environmental chemistry and author and co-author of around 20 scientific articles in peer-reviewed journals. He is also a member of the Geo-Biodiversity and Natural Heritage Laboratory (GEOPAC) at ISR, a member of the management board of Mohammed V University and president of the association of UM5 laureates.

Hind El Bouzaidi is a member of GEOPAC, Mohammed V University, Rabat.

Wassim El-Hajj is an associate professor of Computer Science at AUB. His research focuses on the areas of wireless communication, network security and machine learning. His research activities have resulted in over 90 publications in reputable journals. He is a frequent speaker and technical reviewer at leading international conferences.

Yamina El Kirat El Allame obtained a PhD in Minority Identities, Languages and Cultures. She is an international advisor and consultant in the field of higher education. She is the coordinator of the AMAS project Erasmus Inter Africa Mobility in Morocco. El Kirat is also a coordinator of the Culture, Language, Education, Migration and Society Research Laboratory and of the Studies in Language and Society doctoral program at the Mohammed V University in Rabat in Morocco. For over 25 years, El Kirat has taught and conducted research on the disappearance of languages, cultural representations and attitudes, the analysis of political discourses and minority cultures, languages and identities. She is currently the coordinator of the UniMed Network on Migration.

Faiza El Mejjad obtained a bachelor's degree in English Studies at USMS Beni Mellal before embarking on a professional bachelor's degree in Tourism Engineering at Cadi-Ayyad University in Marrakech. El Mejjad's passion for the environment, space and territories led her to opt for the UM5R master's program Governance, Land Use Planning, Local and Regional Development and Resource Management and the Rural Tourism Heritage option to stay in touch with sustainability and biodiversity. Her master's thesis was a comparative study of media coverage of biosphere reserves in Morocco and Lebanon.

Mohammed Faekhaoui is a member of GEOPAC, Mohammed V University, Rabat.

Bruno Garnier is a university professor at the University of Corsica. The author of 60 publications, he continues his research with a focus on the right to education and cultural diversity in education. Within the laboratory Places, Identities, eSpaces and Activities (CNRS/University of Corsica), he is responsible for the team Identities, Cultures, Heritage Processes (ICPP).

Guillaume Gillet is a research engineer in agricultural and rural training at the National School of Agricultural Education in Toulouse and an associate researcher at UMR Innovation in Montpellier. He coordinates the training of teachers in agricultural equipment sciences and techniques. He conducts his research on support and training for actors who innovate in agriculture with a view to the agroecological transition.

Fatimazahra Hafiane is a member of the GEOPAC, Mohammed V University, Rabat.

Moustapha Itani holds an MSc in Ecosystem Management from AUB. After graduating, he became involved in many projects at the Nature Conservation Center at AUB. Moustapha is currently pursuing his doctoral research assessing the ecological impacts of pastoralism on high mountain landscapes as part of the Interdisciplinary Environmental Science program at the University of Helsinki.

Salma Itsmaïl is a doctoral student in Education Sciences at the University of Corsica and a member of the Places, Identities, Spaces and Activities laboratory (CNRS/University of Corsica). She works on the argan tree biosphere reserve.

Kawtar Jaber is a member of the Faculty of Sciences at USJ, Beirut.

Rhéa Kahalé is a member of the Faculty of Sciences at USJ, Beirut.

Maya Kouzaiha is a member of the Faculty of Sciences at USJ, Beirut.

Jean-Marc Lange is a professor of Education and Training Sciences at the University of Montpellier. His research, carried out from a curricular point of view, aims at the development of coherent and acceptable curricula for the formal education actors in the fields of transversal education. His work also focuses on the issues and orientation of educational policies and their historical roots. He is the author of numerous publications and co-authored the *Dictionnaire critique des enjeux et concepts des "éducations à"* (Critical Dictionary of the Issues and Concepts of "Education For") with Angela Barthes and Nicole Tutiaux-Guillon, published by L'Harmattan in 2017.

Nadia Machouri is a professor at the Mohammed V University in Rabat and specializes in Physical Geography. She is the author and co-author of around 40 publications in national and international journals and doctoral training coordinator in the field of Environmental Management and Sustainable Development. She is also a member of the Center for Studies and Research on Man, Spaces and Societies, having participated in eight national and international research projects.

Roser Maneja is the coordinator of the EduBioMed project. She is a doctor of Environmental Sciences (UAB 2011) and currently Deputy Director of Research and Senior Researcher at the Forest Science and Technology Center of Catalonia (CTFC). She is associate professor in the Geography department of UAB and a member of the Applied Geography Research Group and Visiting Professor at Centro de Investigaciones en Geografía Ambiental (CIGA), Mexico.

Alejandro Marcos Valls is a PhD candidate at ICTA at UAB. He holds an MSc from Uppsala University in Sustainable Development (2014) and two bachelor's degrees, one in Environmental Science (2010) and the other in Media and Communication (2011), both from the University of Barcelona. His current research focuses on integrated scenarios for island economies relying on a metabolic approach to cope with complexity in decision making and to promote participation and deliberation. Alejandro Marcos-Valls acknowledges having received financial support from the Spanish Ministry of Science, Innovation and Universities through the "Maria de Maeztu" Program for Units of Excellence (CEX2019-000940-M).

Anthoni Mas-Ponce carries out research activities mainly related to the monitoring of the ecological condition of Mediterranean river ecosystems and global change. He is currently preparing his PhD at the ICTA at UAB on the creation of new ecological indicators to assess the effects of global change in Mediterranean river basins.

Wahiba Moubhir is an accredited higher education professor in the department of Languages and Human Sciences at the Ecole Normale Supérieure in Marrakech. She is a permanent member of the Laboratory for Studies on Resources, Mobility and Attractiveness in the Faculty of Letters and Human Sciences of Marrakech and an associate member of the Center for Studies, Evaluation and Educational Research (CEERP) of the Cadi-Ayyad University of Marrakech.

Ahmed Mouhyiddine is a Doctor of Geography at Ibn Tofail University, Kenitra (Morocco), and in the Faculty of Letters and Human Sciences, Mohammed V University, Rabat.

Reda Nacer obtained a degree in Geography at the USMBA and a diploma in Business Management before embarking on a master's degree at the UM5R in Governance, Regional Planning, Local and Regional Development and Resource Management. His master's thesis focused on the intercontinental biosphere reserve of the Mediterranean and the subject of wood energy in the province of Chefchaouen.

Salma Nashabe Talhouk is a lecturer at the department of Landscape Design and Ecosystem Management at AUB. His work focuses on the promotion of community-based natural resource management; it explores digital technologies for nature conservation, highlights cultural ecosystem services and looks at the use of native and ecologically adapted plants in towns.

Nivine Nasrallah holds an MSc in Phyto-ecology from the Faculty of Sciences of the Lebanese University. After graduating, she worked as an intern, writing papers and articles at the Nature Conservation Center of AUB. Nivine is currently pursuing a second master's degree in plant protection at the Faculty of Agronomy and Veterinary Medicine of the Lebanese University.

Ken Reyna is the director of the Regional Natural Park and coordinator of the Mont Ventoux biosphere reserve, France.

Bruno Romagny is an economist, a research director (HDR) at the Institut de recherche pour le développement (IRD, Research Institute for Sustainable Development) and a member of the Population–Environment–Development Laboratory (LPED), UMR 151, at the University of Aix-Marseille. Together with Saïd Boujrouf, he co-directed the LMI MédiTer (2016–2021). The issues of access to and uses and modes of appropriation of renewable resources, as well as the difficulties raised by their concerted management and their valorization at a local scale, constitute the heart of his work.

Maria Carmen Romera-Puga is a socio-environmental science researcher interested in the rural commons, community protected and conserved areas, and political ecology issues in the Mediterranean. She is a PhD student at the ICTA at UAB (LASEG research group) and works on ICCAs, *agdals*, inclusive environmental governance and biosphere reserves in Morocco.

Rashid Sammoudi is a PhD research engineer at the Scientific Institute of Mohammed V University of Rabat, a specialist in ecology and environment and the management of natural resources and author and co-author of 10 Scopus scientific articles. A forest engineer and former head of water and forests, he has 20 years of experience in the management and valorization of Moroccan forest resources.

Sònia Sànchez-Mateo holds a doctorate in Environmental Sciences (UAB 2010) and a degree in Biology (UAB 2001). She is a researcher at the ICTA at UAB, a member of the Applied Geography Research Group at UAB and technical manager of Fundació Rivus, a public foundation promoting research, environmental education and dissemination for the conservation of river systems. She also coordinates the Observatori Rivus project, a long-term monitoring project that assesses socio-ecological indicators in Mediterranean river basins. Her expertise concerns riparian vegetation, water quality indicators, protected natural areas, global change and biodiversity.

Melki Slimani holds a doctorate in Education from the University of Montpellier and the Higher Institute of Education and Continuing Education in Tunis. His research focuses on the political dimension of environmental and development issues (*questions environnementales et de développement* (QEDs)) in the curricula. He teaches educational sciences at ISEAH, University of Kairouan, Sbeïtla, Tunisia.

Introduction

This book, *Biosphere Reserves and Sustainable Development Goals*, is organized in two volumes. The first volume specifically addresses scientific issues and educational practices in the Mediterranean region, whereas the second volume focuses on tensions, processes and governance. Case studies examine reserves on the southern, eastern and northern shores of the Mediterranean.

For the first time in the history of the panels of experts in charge of regularly assessing and synthesizing scientific knowledge on climate change and the erosion of biodiversity, approximately 50 Intergovernmental Panel on Climate Change (IPCC) and Intergovernmental Platform for Biodiversity and Ecosystem Services (IPBES) world specialists published a joint report[1] on June 10, 2021. In this document, they stated the imperative need to jointly address these two major phenomena, indicators of the different facets of the "great acceleration" of global change after the Anthropocene (Beau and Larrère 2018), in an attempt to provide responses commensurate with threats and challenges.

The interactions between climate and biodiversity have been more or less known for a long time, but these two areas are still too often treated separately, from the point of view of both research and public policies, or through the "greening" strategies in the activities of large private firms. In this report, researchers show that certain solutions presented as apparently good for the climate – such as single-species plantations of trees popular in carbon offsetting processes – can be harmful for biodiversity, the opposite being rarer. They remind us of the need to take into account these complex interactions at all the possible levels of action.

Introduction written by Angela BARTHES, Catherine CIBIEN and Bruno ROMAGNY.

1 See: https://www.actu-environnement.com/media/pdf/news-37685-rapport-atelier-giec-ipbes-climat.

In France, after being contacted by the municipality of Grande-Synthe in the north of the country (a municipality particularly exposed to climate change), the state council[2] set a new ultimatum to the public authorities. The government had a deadline of March 31, 2022, for implementing all the "useful" measures – without the specification of which – in order to bend the curve of greenhouse gas emissions and meet its goals. Although France has committed itself to reducing emissions by 40% by 2030 compared to 1990 levels and to reaching "carbon neutrality" by 2050, it is not taking concrete measures. Henceforth, administrative judges can control the action or inaction of the State in comparison with the recommended trajectories for the reduction of greenhouse gas emissions. This legal decision, the first of its kind in France, was taken after the High Council for the Climate, an independent body created at the end of 2018 and responsible for shedding light on the government's policies in this area, published a report[3] pointing out that the State was not making sufficient efforts to reach the 2030 goals. This was particularly relevant as the European Union had just raised its own climate goals: it had undertaken to reduce its net emissions by at least 55% by 2030 against the previous 40%, which will require an extra effort on the part of France.

Finance must become a key player in the ecological transition, accompanying its various dimensions (energy, agriculture, food, transport, etc.), all the more so given the fact that the banking sector is dependent on fossil fuel industries as never before. This is clearly shown by a study[4] published by a collective of associations (Friends of the Earth, the Rousseau Institute and Reclaim Finance). The study's abstract quotes the Banking On Climate Chaos 2021 report[5] according to which the 60 largest banks in the world granted EUR 3,393 billion financing to companies in the fossil fuel sector between 2016 and 2020. The in-depth study carried out by this collective of associations shows that the 11 main banks in the euro zone have accumulated a stock of over EUR 530 billion in assets related to fossil fuels, that is, 95% of their equity capital. Since January 2020, BNP Paribas, Société Générale, CréditAgricole and Natixis have granted over USD 17 billion in financing to the 30 most important companies in the world in terms of shale oil and gas exploitation. In this context, the devaluation of "fossil fuel assets"[6] held by the banks, which will

2 This is the body that advises the government on the preparation of bills and decrees. It is also the supreme administrative judge that settles the disputes relating to administrative acts.

3 See: https://www.hautconseilclimat.fr/wp-content/uploads/2021/06/HCC-rappport-annuel-2021.pdf.

4 See: https://gael-giraud.fr/actifs-fossiles-les-nouveaux-subprimes-notre-rapport/.

5 See: https://www.ran.org/bankingonclimatechaos2021/.

6 "Fossil fuel assets" are those contributing to the financing of exploration, exploitation and distribution (including transport, refining, etc.) for oil, gas and coal and to the production of electricity from these resources.

accompany the inevitable ecological transition, could produce significant turbulence or even generate a new global finance crisis. We would find ourselves in a situation comparable to the subprime crisis, when in 2008 the banking sector refused to open its eyes to the upcoming catastrophe and detonated a situation which could have been avoided, resulting in numerous bankruptcies and heavy socio-economic and human consequences.

While scientific warnings are old (Club of Rome report from 1972, etc.), those concerning the interactions between climate change and the rapid erosion of biodiversity have been taking shape for two decades, to the point of describing the current situation as the sixth major mass extinction crisis (Thomas et al. 2004). As the economist Laurent Eloi (2011) has pointed out:

> Our ecological crises reveal a paradox of knowledge and action: the considerable progress in environmental science over the past two decades has brought ever worse news about the state of ecosystems. [...] Even if the natural and physical sciences alert us – by pointing out significant areas of uncertainty – about the reality of ecological crises, they do not provide us with the means to transform attitudes and behaviors in human societies, societies which are responsible for planetary environmental change, behaviors and attitudes alone being capable of influencing its course. [...] In more provocative terms, one could say that, as far as environmental matters are concerned, the social sciences and humanities hold the key to solutions to the problems revealed by the hard sciences. This explains the necessary articulation between the two fields, if ecology does not want to be reduced to an ever more exact science of the contemplation of disasters.

Since the 1990s, various policies described as fostering "sustainable development" have supported mechanisms formalizing the links between the necessary management of biodiversity (Blandin 2009; Rockström et al. 2009) and local development, which must not only be inclusive and favor the reduction of social and territorial inequalities but also as neutral as possible in terms of greenhouse gas emissions. Among these systems, protected areas constitute one of the pillars of national and international nature conservation policies. To be effective, the size of protected areas must globally increase – for both terrestrial and marine ecosystems – and their interconnection has to be improved. However, it is also in terms of multi-stakeholder and multi-level governance that these reserves are expected to truly fulfill their role as an "open-air laboratory" during the ecological transition.

Recent developments systematically remind us of the importance of scientific approaches integrated with the social development needs from all globalized territories, as well as the need to articulate management modes in coordination with the economic and academic spheres. They also highlight the need to increase the engagement from citizens in all the available educational spheres (Cibien 2006).

In one of their last reports[7], experts from the United Nations Environment Programme (UNEP) and the Blue Plan emphasized that the rapidity of climate change in the Mediterranean basin is higher than the global trends. The Mediterranean has warmed by 1.5°C since the pre-industrial era, on average 20% faster than in the rest of the globe. This is what the recent synthesis of several hundred scientific studies has shown regarding the catastrophic consequences of global warming for the inhabitants of this world region, particularly when it comes to the supply of fresh water. By 2040, the Mediterranean region is expected to experience a temperature 2.2°C higher than at the end of the 19th century and locally 3.8°C higher by 2100 if serious mitigation measures are not taken expeditiously. "As a generalist diplomat, I am struck to learn that the Mediterranean is the second most impacted region in the world after the Arctic", summarized the Egyptian Nasser Kamel[8], Secretary General of the Union for the Mediterranean. "I understand that the state of the Arctic is important for the planet, but there are 500 million of us around this small lake..."

As elsewhere on the planet, the impact of climate change in the Mediterranean region has been exacerbated by other environmental problems relating to dynamics which have been underway for several decades, often supported by the public authorities. These phenomena have involved changes in land use, in particular those related to the rise of urbanization and mass tourism concentrated on the coast, and the intensification of agriculture, overfishing, declining biodiversity, soil degradation, desertification and pollution (atmospheric and aquatic environments). At the same time, these phenomena have led to increased competition between local actors and investors – sometimes foreign – for the access to resources related to water, soil (including beach sand)[9], forestry, fodder, etc.

When associated with demographic and socio-economic factors, climate change impacts water resources in various ways. Declining precipitation reduces surface and groundwater recharge. Extreme events (high floods, prolonged droughts)

7 UNEP/Mediterranean Action Plan and Blue Plan (2020). State of the Environment and Development in Mediterranean: Key messages [Online]. Available at: https://planbleu.org/soed/.

8 See: https://www.lemonde.fr/planete/article/2019/10/11/changement-climatique-les-scientifiques-s-inquietent-des-effets-sur-le-bassin-mediterraneen_6015045_3244.html.

9 See: https://wedocs.unep.org/bitstream/handle/20.500.11822/28163/SandSust.pdf.

accelerate soil erosion and threaten many human lives, as do recurrent heat wave[10] episodes. Water quality also suffers an impact, in particular due to the deterioration of wetlands and the salinization of groundwater in coastal aquifers.

The agricultural sector in the Mediterranean being the largest water consumer, the risks of conflicts between users have increased. In connection with demographic growth and the expansion of metropolitan areas, food demand is bound to increase accordingly (both at the national level and for export), which will automatically increase the demand for irrigation. Technical solutions in terms of water saving, such as dripping (Kuper et al. 2017), only postpone or delay the overexploitation of underground aquifers and create other difficulties (soil pollution by plastics abandoned in the open field, etc.). Significant adaptations are expected to modify agricultural[11] practices, defend the traditional Mediterranean diet, support localized agricultural production systems or even reduce waste and encourage a change in diets in line with the functioning of global food markets.

In the Mediterranean, terrestrial biodiversity is undergoing many changes. In the countries of the northern shore (especially coastal ones), urbanization has suppressed or fragmented many ecosystems. Due to the decline in agricultural and agro-pastoral systems, wooded areas are increasing at the expense of these two sectors. For their part, the semi-natural ecosystems of the countries along the southern and eastern shores are threatened with fragmentation or disappearance, notably due to urbanization, deforestation, the overexploitation of wood (for heating, construction) and overgrazing. According to the mentioned UNEP report, adaptation options at the level of terrestrial biodiversity include the preservation of "natural flow variability in Mediterranean rivers and the protection of riparian zones, reduction of water abstraction, modified silvicultural practices, and the promotion of climate-wise landscape connectivity".

---

10 For over a decade, each new year has broken the temperature records of the previous one in many countries on all continents. The summer of 2021 caused alarm with the appearance of a "heat dome" in Western Canada which resulted in the hottest day ever recorded in this country on June 30, 2021, with a 49.6°C temperature recorded at Lytton. This was one of the coldest territories on Earth showing the usual temperatures of the Middle East! As if to reinforce the dramatic dimension of the phenomenon and adding to its symbolism, the following day, the village of Lytton was almost completely destroyed by the flames of fires caused by this unprecedented heatwave episode. Other large fires also marked the summer of 2021 around the Mediterranean, specifically in Turkey, Greece, Algeria, France, etc.

11 Towards a productive agriculture capable of feeding the planet differently, in a less carbon-intensive manner, less chemical, less intensive and more respectful of humans, soils and biodiversity and more water-efficient.

Furthermore, crucial socio-economic factors are currently hampering sustainable development in Mediterranean countries. In the first place, there is persistent poverty, for example, in certain marginal rural regions in Northern Africa and in the urban and periurban suburbs which accommodate the most disadvantaged populations (unemployed, landless peasants, migrants, etc.). These pockets of poverty are related to inequalities and the imbalance between men and women. In addition, cultural dimensions also have to be taken into account to ensure the success of climate adaptation and environmental resilience policies in the heterogeneous multicultural framework of the Mediterranean basin. These policies, also aiming to support the most vulnerable local communities, must imperatively take into account social issues such as justice, equality, education, the fight against poverty, social inclusion and redistribution.

Finally, at the scientific level and with a view to meeting the main challenges in the Mediterranean, UNEP and Blue Plan experts insist on the need to quickly fill in the gaps between countries in terms of data and knowledge. In particular, they aim to promote the development of high-level climate services. This essentially points to early warning systems: "Increased research is required for short- and medium-term projections, as well as large-scale programs at the Mediterranean level to face impending challenges."

Since 1971, the UNESCO Man and the Biosphere Programme, known by the abbreviation MAB, has fostered different articulation principles between the political, scientific and academic spheres in various territories. Theoretically, the declared goal is to reconcile sustainable development at the regional level with the protection of the environment and, more specifically, the conservation of biodiversity, with respect for the cultural values of all. Biosphere reserves, which are formalized by UNESCO, are considered privileged territorial spaces for experimenting with operating methods specifically adapted to the stated goals and cross-cutting issues.

Today, the UN biosphere reserve system supports the 2030 Agenda and its 17 Sustainable Development Goals (SDGs). It encourages the development of interdisciplinary research and relies on its worldwide network to spread its experiences, its approaches and know-how. The different global and local political scales are articulated, with various consequences on the reconfiguration of local political arenas; on specific modes of development regarding the renewal of bonds between knowledge, powers and institutions and renovated relationships between the scientific and educational spheres; and on the governance of territories.

Responding to a multidisciplinary requirement and to a citizen–actor–researcher dialogue, this work addresses the question of the articulations and tensions between biosphere reserves and SDGs in the Mediterranean area from the perspective of complexity and multi-referentiality. It was conducted within the framework of the

Erasmus+ EduBioMed[12] (2018–2021) European program, led by the Autonomous University of Barcelona (UAB). This program also engages five other universities: the University of Aix-Marseille in France, Mohammad V in Rabat and Cadi-Ayyad in Marrakech in Morocco, and Saint-Joseph University and the American University in Beirut in Lebanon. It is also established a partnership with several biosphere reserves in France (Mont Ventoux), Spain (Montseny), Morocco (the Arganeraie and the Intercontinental Reserve of the Mediterranean) and Lebanon (Jabal Moussa, Chouf), as well as with the MAB France and UniMed networks.

The book is structured in four distinct and complementary parts. The first part focuses on the multidisciplinary scientific issues of biosphere reserves as part of the implementation of the SDGs in the Mediterranean. The second part focuses on educational and civic prescriptions and practices relating to biosphere reserves. The third part (Volume 2) guides reflection on the governance processes in climate change on both sides of the Mediterranean. Finally, the last part (Volume 2) engages the actors to express themselves on local issues specific to the south of the Mediterranean. This book has the particularity of resolutely taking part in the cross-dynamics of Mediterranean biosphere reserves and is enriched by the points of view from Lebanese, Tunisian, Moroccan, French and Catalan actors involved in the field and in the academic sphere.

This book intends to make its contribution to the specificities of the Mediterranean biosphere reserves, analyzing the articulations and tensions between the development of hinterland territories and environmental protection. It questions to what extent the challenges of the erosion of biodiversity and governance difficulties encountered in the Mediterranean biosphere reserves are actually taken into account. It also contributes to formalizing the links between academic research, citizens and biosphere reserve managers without concealing issues such as knowledge conflicts, citizen involvement and educational realities in a problematic framework which ends up being more complex than suspected at first sight.

References

Beau, R. and Larrère, C. (eds) (2018). *Penser l'Anthropocène*. Presses de Sciences Po, Paris.

Blandin, P. (2009). *De la protection de la nature au pilotage de la biodiversité*. Quae, Versailles.

Cibien, C. (2006). Actualités de la recherche. Les Réserves de biosphère : des lieux de collaboration entre chercheurs et gestionnaires en faveur de la biodiversité. *Natures Sciences Sociétés*, 14, 84–90.

12 See: https://www.edubiomed.eu/fr/.

Eloi, L. (2011). Économie du développement soutenable. Quelle place pour l'économie dans la science de la soutenabilité ? *OFCE*, 120, 7–12.

Kuper, M., Ameur, F., Hammani, A. (2017). Unravelling the enduring paradox of increased pressure on groundwater through efficient drip irrigation. In *Drip Irrigation for Agriculture: Untold Stories of Efficiency, Innovation and Development*, Venot, J.P., Kuper, M., Zwarteveen, M. (eds). Routledge, London.

Rockström, J., Steffen, W., Noone, K., Persson, Å., Chapin III, F.S., Lambin, E.F., Lenton, T.M., Scheffer, M., Folke, C., Schellnhuber, H.J. et al. (2009). A safe operating space for humanity. *Nature*, 461, 472–475.

Thomas, C.D., Cameron, A., Green, R.E., Bakkenes, M. (2004). Extinction risk from climate change. *Nature*, 427, 145–148.

PART 1

Biosphere Reserves and Sustainable Development Goals: Multidisciplinary Scientific Issues

Introduction to Part 1

Considering the "biosphere reserve" as a research object, the goal of this first part of the book is to examine the methods that enable scientists (via pluralist approaches and the need to establish a constructive dialogue between disciplines) to better feed political and citizen debates regarding the transformations to be tackled in many fields in order to achieve the Sustainable Development Goals (SDGs) by 2030. Although the evaluation of the actions implemented marks significant progress, the prospects for all the countries in the Mediterranean basin show that there is still a long way to go in less than a decade. The findings presented by UNEP and Blue Plan experts in the above-mentioned[1] report are worrying:

> Despite notable progress, Mediterranean countries are not on track to achieve and fully implement the goals they have agreed to, including the SDGs and the goals for the good ecological status of the Mediterranean Sea and its coast. Most of the trends observed show developments that are progressing either towards the goals set but at an insufficient pace or unevenly from one country to another or are moving away from the goal. In 2019, 9 of the 21 Mediterranean countries had not achieved any of the SDGs for 2030, and the maximum number of SDGs achieved per country was 2 (out of 17). Eleven SDGs remain to be achieved in all Mediterranean countries, including SDG 13 "fight against climate change", SDG 14 "life below water" and SDG 15 "life on land". Regarding SDG 2 on "hunger", SDG 5 on "gender equality", SDG 11 on "sustainable cities and communities" and SDG 14 on "life below water", none of the Mediterranean countries shows a trend making it possible to achieve

Introduction written by Bruno ROMAGNY.

1 UNEP/Mediterranean Action Plan and Blue Plan (2020). State of the Environment and Development in Mediterranean: Key Messages. Nairobi. Available at: https://planbleu.org/soed/.

these goals by 2030. Major changes in production and consumption patterns need to be urgently made in order to move decisively towards inclusive sustainable development, with a focus on climate change concerns, biodiversity protection and ecosystem restoration, pollution prevention and the circular economy. The transition towards a blue/green economy is a challenge in the region which requires the redirection of financing from polluting activities towards sustainable activities.

Given this significant delay in achieving the SDGs in the Mediterranean, we can reflect on the contributions of scientists to help accelerate awareness at the regional level, and above all, to suggest actions to be taken now in all sectors of the economy and in all areas of life. In general, scientists are expected to alert the public and propose solutions adapted to contexts, while politicians are expected to act in accordance with democratic rules and within the framework of their elective mandates. However, the reality of the relationship between science and politics does not necessarily match this role distribution, as we have recently observed, for example, with the Covid-19 pandemic or the debates on academic freedom in French universities. Democracy is in crisis; this is nothing new. Almost everywhere in the world, democracy suffers from significant abstention and from the rise of nationalist, racist and xenophobic discourse multiplied by social networks and by the new manipulators of information, something rarely done for benevolent purposes.

The absence of strong public policies, proactive in character and coordinated to favor the ecological transition, might give the impression that politicians remain deaf to the scientists' warnings. Things are far from being simple, as the analyses by Sélim Louafi[2] (political science researcher at CIRAD) have shown:

> Whether on issues related to climate change or biodiversity – characterized by high degrees of scientific complexity and uncertainty and by deep divergences in the values and interests between the actors – science is strongly called upon to help define the problems, develop solutions or evaluate the policies implemented. [...] Hardly a week goes by without a new study reminding us of the seriousness of environmental issues. At the same time, the observation is clear when it comes to translating these alarming reports into strong political action: nothing is happening! While politicians are often blamed, what can be said about scientists themselves? What if the problem of inaction was also derived from science and its difficulty of adequately acknowledging the political dimensions of its objects and knowledge production practices?

2 See: https://theconversation.com/les-scientifiques-alertent-mais-les-politiques-ne-font-rien-est-ce-vraiment-si-simple-158205.

Sélim Louafi identifies a number of received ideas on the complex relations between science and politics. The underlying concept behind these clichés is that science produces objective and neutral knowledge, "ready for use" by politicians unable to grasp it, obsessed as they are with the next elections and equipped with a short-term vision. Establishing a scientific consensus before approaching decision-makers is far from easy. On the other hand, "if science were called upon to produce politically relevant knowledge, it could never become prescriptive on the decisions to be taken. This watchword constitutes an ever-powerful mantra within the IPBES or the IPCC". Is it reasonable to maintain this role distribution, this distinction between scientific and political authorities while global changes keep accelerating? Reflective work should be carried out on the intricacies between these two spheres:

> This reflective work, from the individual level to that of scientific institutions, cannot be limited to a simple alert or guarantee function, as classically done by ethics committees, for example. It involves a refusal to grant to certain disciplines the monopoly over objects (for example, biodiversity only for ecologists or GMOs for geneticists) and a change of posture in which research, following a pragmatist approach, is designed as a collective and collaborative investigation with a wide range of non-academic actors influenced or concerned by the issues addressed (transdisciplinarity). By placing collective learning at the center of this renewed link with societal components, the question comes down to building new action-focused knowledge capable of establishing a more effective bond between the world of research and that of decision-making.

The texts brought together in this first part contribute, each in their own way, to the elements for debate we have just outlined. To begin with, Meriem Bouamrane and Didier Babin offer a historical reflection on the MAB Programme and its innovative nature and show to what extent the biosphere reserves carry within them the "seeds" to be sown to prepare the "next world". In the authors' words, "one of the key lessons of the MAB Programme is that it is also, and above all, outside protected areas that one needs to invest in for the design and implementation of a solidary ecological transition. It is not only the biological or physical environment that preserves biodiversity, but also the social and economic systems which are connected to biodiversity and ecosystemic services". They reposition biosphere reserves in sustainable development policies and give us a systemic reading of current transformations. This reflection is continued, from another angle, in the contribution by Magda Bou Dagher Kharrat, Éliane Bou Dagher and Rhéa Kahalé entitled "Humans and Nature: A Story to be Rewritten". Going far back in the history of human/nature relationships, the authors end up reflecting on the contribution of the biosphere reserve system in the Mediterranean to the *nexus* sought among the different SDGs.

Angela Barthes, Bruno Romagny, Jean-Marc Lange, Lahoucine Amzil, Roser Maneja, Mohammed Aderghal and Véronique Chalando complete this backdrop by proposing a study of multicultural social representations regarding biosphere reserves. Based on a quantitative analysis of the data collected from several groups of students associated with the EduBioMed program, they reveal the complexity of transferring models and standards to the Mediterranean region. For their part, Moustapha Itani, Salma Nashabe Talhouk, Wassim El-Hajj, Nivine Nasrallah and Hannah Abou Fakher explain to us the challenges and opportunities of collaborative research on biosphere reserves in the Mediterranean, as seen from Lebanon. This country, partner of the EduBioMed program and having hosted part of the research team in the field at the end of April 2019, has been going through a political and economic crisis for many months and was cruelly marked by the explosions at the port of Beirut in August 2020.

Focusing on a case study in Hautes-Alpes (France), Mikaël Chambru and Cécilia Claeys strive to grasp the socio-environmental challenges of the touristic development of scientific culture in the context of multi-labeled protected areas. How does the touristic development of scientific culture in mountain territories reveal the pre-existing tensions between democratic, cultural and environmental issues as a response to the injunction of the ecological transition? The authors' analysis is structured around three hypotheses relating to the forms of transition brought about by the touristic development of scientific culture from a socio-political, cultural and environmental perspective.

1

Man and the Biosphere: A Precursory Program for the Next World

1.1. 1971–2021, the beginnings of sustainable development

Over 50 years ago, following the Biosphere Conference held in Paris in September 1968, visionary men and women created an intergovernmental scientific program, the Man and the Biosphere (MAB) Programme, within UNESCO whose mission is still urgently relevant. From that very moment, the program focused on interdisciplinary scientific approaches regarding the relationships between humans and ecosystems. This program aimed to promote solutions to reconcile the conservation of biodiversity with the economic, social and cultural development of human populations. The MAB Programme proclaimed that the conservation of natural resources should go hand in hand with their rational use and that we could not content ourselves with putting nature "under a glass dome" in strictly protected areas. This vision of an ecology of reconciliation, the preliminaries of sustainable development, resulted in the creation of exemplary sites around the world, recognized by UNESCO under the name "biosphere reserves". These are territories where the inhabitants have a common project aimed at flourishing and prospering in harmonious interaction with the biosphere in the long term. Much progress has been made since the creation of the first biosphere reserve in 1976. Today, the global network[1] has 714 sites in 129 countries, including 21 transboundary biosphere reserves hosting over 275 million people.

They are "sites for experimentation, demonstration, practice, confrontation with reality and dissemination of sustainable development approaches, know-how and practices… Good practices in farming, fishing or forestry management, sustainable

Chapter written by Meriem BOUAMRANE and Didier BABIN.

1 See: https://en.unesco.org/biosphere/wnbr.

architecture, biodiversity conservation, urban planning, education for sustainable development, planning, research on and organization of the participation of actors in the field, youth engagement… which can be deployed beyond the limits of the biosphere reserves. They are key to achieving the sustainable development goals defined by the United Nations in 2015 as quickly as possible" (Babin 2021).

The ambition of the MAB Programme and its global network of biosphere reserves is to contribute to the profound transformations required to meet current and future socio-ecological challenges. The aim is to participate in the creation of "a world where people are aware of their common future and their interaction with our planet and act collectively and responsibly to build prosperous societies in harmony with the biosphere" (excerpt from the Lima action plan adopted in 2006).

According to reports published by the Intergovernmental Science-Policy Platform on Biodiversity and Ecosystem Services (IPBES) in 2019 (Díaz et al. 2019), the main global drivers of biodiversity loss are climate change, invasive species and the overexploitation of natural resources, as well as pollution and urbanization. Land degradation occurs in all parts of the terrestrial world and can take many forms. Combating land degradation and restoring degraded land is an urgent priority to achieve the Sustainable Development Goals (SDGs) contained in the United Nations agenda for 2030.

Unfortunately, we are witnessing a rapid decline in biodiversity, which threatens both nature and people. The widespread degradation of ecosystems primarily affects the poorest and most vulnerable populations. At a global scale and according to an anthropocentric vision, not only does it jeopardize the contributions of nature to human societies but also economies, means of subsistence, food and nutritional security, cultural and spiritual diversity and the quality of life of individuals. These ecosystem services, dependent on biodiversity, include the provision of drinking water, food and fiber; soil fertility; maintenance of wild and domestic genetic diversity; climate regulation; and recreational and aesthetic values, among others.

Nowadays, we have to accurately assess the scope of this pioneering program that goes well beyond the conservation of nature or its preservation and aims to be more ambitious than the 30% goal for protected areas by 2030 under discussion as part of the negotiation of the post-2020 agreement on biodiversity. This goal and others were discussed in Kunming during the 15th Conference of the Parties (COP 15) at the Convention on Biological Diversity (CBD) held in China in October 2021. The crisis we are going through has had a strong impact on us: it becomes urgent to reconcile 100% of human beings with the other species of the living world, and this everywhere on Earth, for current and future generations.

1.2. Making sure no one is left behind

Sustainable development includes the maintenance of terrestrial/marine mosaics, a mixture of natural, rural and urban ecosystems where local communities and other stakeholders actively encourage governance and management authorities to guide decision-making processes and to generate options for the environmental, economic and social wellbeing of present generations, without impeding that of future generations. Sustainable development is based on the assumption that the health of ecosystems and the richness of biodiversity are related to policies and practices which create opportunities for humans to improve their social, economic and cultural conditions. Conserving biodiversity, maintaining the resilience of ecosystems or their restoration and safeguarding the diversity of life on our planet are fundamental to a global transformation towards sustainable and resilient societies. The loss of biodiversity also disproportionately affects the most vulnerable, exacerbating inequalities.

In many societies, women embody specific knowledge on biodiversity and apply it in a sustainable manner. However, their role in managing biodiversity and decision-making processes may be incorrectly recognized or capitalized on. Gender equality issues are specifically targeted by SDG 5 but women are directly affected by the achievement of several other SDGs (1, 2 3, 4, 6, 10, etc.). Biodiversity and cultural diversity are closely related. Accumulated across generations and renewed by each new generation, indigenous and local knowledge is also an integral part of a cultural complex comprising language, classification systems, resource use practices, social interactions, rituals and spirituality. These unique ways of knowing are essential facets of global cultural diversity and provide a basis for sustainable development and a locally appropriate way of life.

In addition, for several years, the concerns of youngsters have been expressed in a consistent way in the face of the significant deterioration risks in their living conditions for their future. Young people's voices are increasingly heard in the media and by politicians, businesses and local communities, but taking into account their interests is still weak with regard to decisions which threaten their future. Supporting and promoting the contribution of women, youth and indigenous peoples as agents of change is essential to inform us about the factors degrading biodiversity and to reverse this alarming trend.

While stopping the loss of biodiversity mainly concerns SDGs 14 and 15 (life below water and life on land), biodiversity is also strongly related to all the other SDGs. The implementation of the SDGs must be inclusive and based on a profound transformation in the roles, actions and relationships of people with nature. Awareness and appreciation of the diverse values of biodiversity need to be further amplified and integrated into the 2030 Agenda in order to transform human behavior in favor of the conservation and sustainable use of biodiversity. A diverse world gives us the flexibility to adapt to local and global changes, including climate change. Biodiversity pervades most of the SDGs, and its loss poses a threat to both global security and peace.

1.3. Identification of gaps, risks and challenges

For numerous actors, the designation and management of protected areas are the cornerstone of biodiversity conservation. Despite an increase in the total number and area of protected areas worldwide, biodiversity continues to decline dramatically, in part due to the limited resources to properly maintain and manage these areas and/or to enforce relevant legal frameworks.

One of the key lessons of the MAB Programme is that it is also, and above all, outside protected areas that we need to invest in for the design and implementation of a solidary ecological transition. It is not only the biological or physical environment that preserves biodiversity but also the social and economic systems which are connected to biodiversity and ecosystemic services.

The MAB Programme supports Member States and their populations in their efforts to halt the loss of biodiversity by understanding, appreciating, safeguarding and using biodiversity in a sustainable manner. Biosphere reserves offer an integrated or connected approach to landscapes and seascapes, including the essential conservation planning to ensure suitable habitats for living species. Living in and managing a biosphere reserve reconciles conservation and development goals within the same space, while promoting the convergence of the short-, medium- and long-term interests of the actors involved and their descendants.

Biodiversity conservation and sustainable development are still goals to be achieved in a context of great uncertainty. To address them, we need to collectively promote and explore social and ecological interdependencies and solidarities by means of participatory processes and scenarios. Despite the unparalleled availability of knowledge and data, enough to put an end to the erosion of biodiversity, there still are gaps in the trends and drivers of evolution of many ecosystems and species subject to anthropogenic pressures. Interdisciplinary research and data sharing are needed to fill these gaps and to identify policy responses which will take into account different forms of knowledge, economic paradigms and cultural norms. A better approach to all the available knowledge must be found, including citizen or participatory science, as well as summoning indigenous and local knowledge.

These different forms of knowledge and skills are implemented on biosphere reserves. Embracing change and social learning is a central issue to facilitate the future institutional, technical and social innovations required to adapt to local or global changes. The magnitude and intensification of the erosion of biodiversity show that delaying action will only increase costs in the future, leading to an even greater loss of benefits, with a higher impact on people's means of support.

1.4. Valuable lessons learned from the transformation towards sustainable and resilient societies

Resilience is reinforced by participation. It is through participation and meaningful engagement that individuals in a group become members of a community. These interactions facilitate the flow of information and knowledge needed to meet the challenges. Based on the vision of the MAB Programme, it is necessary to foster the commitment and actions of a diversity of stakeholders, decentralized governance, regional cooperation and partnerships with the private sector. The action plan adopted in 2016 during the Fourth World Congress of Biosphere Reserves organized in Peru, as well as the Lima declaration, serves as a roadmap for the implementation of the MAB Programme strategy over the period 2015–2025. This roadmap places a strong emphasis on prosperous societies in harmony with the biosphere for reaching the SDGs and implementing the sustainable development program by 2030, both in biosphere reserves and beyond.

The MAB Programme constitutes an important and valuable partner and instrument for research and field experimentation at the service of sustainable development, based on practice and sharing at a global scale. This program leverages lessons learned from science and education in sustainability. It uses modern, open and transparent means to communicate and share information. This is how the biosphere reserves global network encourages not only the dissemination of a certain model of sustainability, but also the sharing of experiences aimed at improving local governance, collaboration and networking, as well as developing effective external partnerships to ensure long-term sustainability. The results of case studies from this network show how biosphere reserves are concrete tools to achieve a global transformation towards sustainable and resilient societies and to implement the SDGs holistically.

1.5. Investments that may affect the building of sustainable and resilient societies

First and foremost, investing in sustainable development in order to build sustainable and resilient societies involves stopping the investment in activities, projects or infrastructures incompatible with strong sustainability, that is, activities which do not take into account the biosphere and future generations. Even though financial resources exist, they are often invested considering short-term interests, representing a minor portion of beneficiaries. While this may be acceptable for private financing, it seems totally incoherent for public financing, which must enable the implementation of public policies advocating the preservation of biodiversity and targeting the SDGs. Many economic sectors are still highly dependent on subsidies harmful to biodiversity and incentives incompatible with the achievement of SDGs as a whole. A significant effort must be made to quickly and publicly identify these harmful subsidies or investments to contemplate their disappearance or transformation so that they become at least compatible (and if possible, positively so) with strong sustainability, acknowledging intra- and intergenerational equity. Over time, this will also require supporting the transition so that harmful activities can be transformed into sustainable and resilient ones.

1.6. Integration of biodiversity within sustainable development policies

A lack of adequate resources and of technical and financial capacities can limit the scaling up of innovative solutions. However, above all, most biodiversity-focused policies are fragmented and aimed at specific or restricted targets. Integrating biodiversity into cross-sectoral development policies, plans and programs contributes to the efforts to reach the SDGs. This is particularly the case for mitigating climate change and for succeeding in the transition towards sustainable development. It is also necessary to take into account current and future environmental migrations, which largely depend on the adaptation strategies implemented in vulnerable regions and on the mitigation of environmental degradation and climate change.

The governance and management of biosphere reserves particularly rely on the crucial role played by knowledge and on learning and the reinforcement of capacities for creating and sustaining dynamic and mutually beneficial interactions between the conservation and development functions within these reserves. Efforts should be redoubled to develop national biodiversity strategies and action plans (*Stratégies et plans d'action nationaux pour la biodiversité* (SPANBs)). Such plans should be based on the co-production of knowledge in close connection with civil society and other user groups, including the private sector, to inform policy decisions and strengthen national policies related to biodiversity where these already exist. Other factors are also at the heart of these SPANBs, such as the importance of multilevel governance (national/local/community) and the support of the regional and international network of biosphere reserves from the MAB Programme.

In line with the holistic approach of the 2030 Agenda, SDGs 14 and 15 should be considered in relation to all the other SDGs. In order to avoid compartmentalized approaches according to which biodiversity issues are addressed at the national level by the ministries in charge of the environment, it could be interesting to promote SPANBs as tools for implementing the SDGs, rather than considering them as independent strategies. In addition, it is essential to identify SDG targets which could directly or indirectly harm biodiversity and ecosystem health. This could support alert mechanisms on these targets and goals, in order to seek solutions and transform these risks into opportunities both for the biosphere and for humans. It is also important to raise awareness about these positive solutions in sectors which are generally unaware of the risks they can pose to biodiversity.

1.7. Policy recommendations to accelerate progress in building sustainable and resilient societies

Attempts at raising awareness on the importance of biodiversity and effective communication around these complex issues are rarely a political priority. The depreciation of natural capital is little integrated into the national accounts. It is essential to better apprehend this dimension and to strengthen the mechanisms that internalize the externalities created by the degradation of the biosphere to responsibly inform decision-making processes. The lack of knowledge on biodiversity and its importance is widespread, biodiversity sometimes being perceived as a simple resource to be exploited, as occurs with the unsustainable management of forests and fisheries or the poaching of wild animals.

Communication on biodiversity is a critical issue that needs to be addressed to reach the SDGs and the goals from the Convention on Biological Diversity (CBD). A major challenge is to draw attention to the importance and urgency of integrating biodiversity in the frame of the 2030 agenda. The question comes down to obtaining the indispensable high-level support, not only via governments and the United Nations system but also from civil society, so as to shed light on the negotiations on the future post-2020 global biodiversity framework and its effective implementation in the field. Another challenge is to develop a powerful common narrative to communicate biodiversity to policy makers and to civil society in view of transforming our relationship with the living world. This narrative should be able to engage key stakeholders in the process, including the most vulnerable actors (youth, women, etc.), local communities, businesses and the private sector, by persuading them to become actors in this transformation towards resilient societies.

1.8. Lessons learned from the Covid-19 crisis and perspectives for biosphere reserves for the next world

Our socio-economic and political systems will only be able to overcome this crisis by calling themselves into question at a core level. The pandemic has revealed our interdependence with the living world. The destruction of biodiversity is mainly derived from the crisis of our relationships with other living species. Humans are responsible for 75% of the alteration of terrestrial ecosystems through their activities (Díaz et al. 2019). However, the MAB global network is rich with many examples of good practice associated with worldviews and cosmologies in which humans do not dominate nature[2]. This reconciliation involves recognizing the diversity of values associated with the living and seeks to reconnect humans to the biosphere by modifying their behavior at the individual level, as well as by promoting collective actions that can shape public[3] action. Take a look at the food systems, for example: while regulation can contribute to positive changes (supply chains, reduction of harmful subsidies, etc.), at the same time, all individuals can work towards more sustainable and health-based diets, which would simultaneously reduce human mortality and the loss of biodiversity. It is these transformations and these united transitions that we must now accelerate in order to be able to continue to live on Earth.

In relation to food security in particular, science/policy interfaces on biodiversity emphasize the existing links between pandemics and the attacks on biodiversity, the trafficking of wild animals, intensive farming, etc. Among other disparities, this signals a major dysfunction in our relationship to the living. Once recovered from the social and economic shock suffered on an individual and planetary scale, will these complex interactions between biodiversity and human activities be sufficiently understood to really consider other forms of relationship with the living world? More than ever, economic recovery plans are trying to take the ecological and, above all, climatic dimensions into account.

It is still far too early to know what will really happen in the months and years to come. However, there is a risk of seeing short-term priorities taking precedence over the profound and necessary transformations to be envisaged in order to prevent new threats of this type. The recovery of economic activities should not result in a further deterioration of living conditions on the planet, in particular for the most vulnerable

2 For example, areas inhabited by Indigenous Peoples Local Communities (IPLCs) represent about 35% of the last areas of the globe with very little human intervention. These peoples have applied principles and indicators such as the land's health, the protection of the country and mutual responsibility, contributing to the improvement and maintenance of biodiversity and of wild and domesticated landscapes in their territories.

3 See: https://theconversation.com/the-four-ecologies-of-lanthropocene-152490.

populations, nor should they further reinforce activities that are harmful to the biosphere, not to mention create new ones. The shock that the implementation of genuine sustainable development could represent for many economic sectors is becoming evident. The ecological and solidarity-based transition is all the more relevant in that, over an acceptable period, it should encourage and support the positive transformations essential for the planet and our common future.

As stated by the philosopher Frédéric Ducarme, "there is only one planet, and not one planet for nature and another for humans" (Ducarme 2020). So far, we have conceived this planet as a place or as a scenery, challenging the local and indigenous representations which have always argued the need for an integrated approach to the relationship between humans and the biosphere. It is time to refer to this Earth where we humans live, or co-live with all non-human forms, whether we perceive its importance or not. This idea of cohabitation, and its implications regarding the joint shaping of a common home, immediately calls for an ethic.

These crises we are going through on a planetary scale have awakened or revealed the values we can rely on to envisage the world that we wish to build for ourselves and the generations to come. The year 2021 was suitable for inaugurating positive transformations at a global scale, when a variety of international meetings were held for discussing biodiversity, the climate, the fight against desertification, the oceans and agri-food systems. Today, it is urgent to take action, as the World Network of Biosphere Reserves has been doing for the past 50 years, and to invest in the construction of solidarity bonds and in the human capacities for transformation and for listening to, understanding and respecting the signs that the other species we live with emit. It is necessary to take full responsibility for no longer destroying the conditions of habitability of our planet Earth and to transmit this sense of responsibility to all generations, present and future ones.

Let us positively support these movements at our level, based on concrete and embodied experiences in biosphere reserves, mobilizing the young for a more resilient future, relying on sustainability sciences and on public and private investments with positive effects for the biosphere and no longer harming it. If we do not act in the current circumstances, how big will the catastrophes have to be to force us to alter the trajectory of humanity on planet Earth?

1.9. References

Babin, D. (2021). 1971 : Lancement du programme MAB. In *La Science au présent 2021. Une année d'actualité scientifique et technique*. Encyclopedia Universalis France, Paris.

Blandin, P. (2009). *De la protection de la nature au pilotage de la biodiversité*. Quae, Versailles.

Cibien, C. (2006). Actualités de la recherche. Les Réserves de biosphère : des lieux de collaboration entre chercheurs et gestionnaires en faveur de la biodiversité. *Natures Sciences Sociétés*, 14, 84–90.

Díaz, S., Settele, J., Brondizio, E.S., Ngo, H.T., Guèze, M., Agard, J., Arneth, A., Balvanera, P., Brauman, K.A., Butchart, S.H.M. et al. (2019). Summary for policymakers of the global assessment report on biodiversity and ecosystem services. Report, IPBES, Bonn.

Ducarme, F. (2020). L'idée de nature est un concept très… culturel. *Le Monde* [Online]. Available at: https://www.lemonde.fr/sciences/article/2020/10/10/frederic-ducarme-l-idee-de-nature-est-un-concept-tres-culturel_6055580_1650684.html.

Rockström, J., Steffen, W., Noone, K., Persson, Å., Chapin III, F.S., Lambin, E.F., Lenton, T.M., Scheffer, M., Folke, C., Schellnhuber, H.J. et al. (2009). A safe operating space for humanity. *Nature*, 461, 472–475.

Thomas, C.D., Cameron, A., Green, R.E., Bakkenes, M. (2004). Extinction risk from climate change. *Nature*, 427, 145–148.

2

Humans and Nature: A Story to be Rewritten

2.1. *Homo sapiens*, a species like the others

There was a time when several human species coexisted on the surface of our planet. Just as many equine[1] species currently co-exist, *Homo sapiens* (whose origin dates back 300,000 years according to the latest discoveries at Jabal Irhoud in Morocco, see Hublin et al. 2017) was contemporary with at least four other human[2] species. These are now all extinct, and *Homo sapiens* ruthlessly acts alone on our planet's surface.

For most of its existence, *Homo sapiens* was a species like the others. In order to survive, they collected and gathered plants, they fished and hunted, and they were sometimes themselves hunted by natural predators. They moved about in accordance with the rhythm of the seasons and followed game which, in turn, followed after the vegetation, depending on climate changes.

2.2. *Homo sapiens*, a nature modifier

Only 12,000 years ago, in the "fertile crescent" area currently corresponding to Iraq, Syria and Lebanon, climatic and environmental conditions become favorable to sedentarization. *Homo sapiens*, tired of hunting large beasts which became rarer and

---

Chapter written by Magda BOU DAGHER KHARRAT, Éliane BOU DAGHER and Rhéa KAHALÉ.

1 Horses (*equus caballus*), donkeys (*equus asinus*) and zebras (*equus quagga*).

2 *Homo neanderthalensis* (who disappeared 28,000 years before our era), Denisova hominins (who disappeared 30,000 years ago), *Homo floresiensis* (who disappeared 12,000 years ago) and *Homo naledi* (who disappeared 30,000 years ago).

rarer, began to domesticate plants and animals. On the one hand, they invented agriculture and, on the other hand, breeding, through the domestication of certain species. They began exploiting and storing the genetic resources present and modifying them by selecting the characteristics which best suited their needs. This same pattern repeated itself in several places in our biosphere, a few thousand years apart. The biosphere offered humans the resources that enabled them to develop and colonize almost the entire planet.

As human society developed and the world's population increased, the relationship between humans and nature changed. Agriculture and breeding gave rise to a set of human practices that would destabilize the Earth system. These practices pushed ecosystems to produce more than they were able to spontaneously supply. Humans thus passed from the status of "consumer" to that of greedy and venal operator. With sedentarization, the sense of private property through land possession and securing triggered multiple conflicts and wars.

The greater the resource scarcity, the higher the risk of conflict (Homer-Dixon 1999). Since the end of World War II, the human population has experienced unprecedented growth. Population growth and the rise in consumption levels led to a scarcity of natural resources accessible to each person, thereby creating a climate of instability and accompanying conflict. When vital resources such as fertile land become scarce due to rapid population growth and imbalanced access to land, resource-poor people will be driven to ecologically marginal areas, such as upland hillsides, areas at risk of desertification and tropical rain forests.

Higher population densities in these areas, along with environmentally unsustainable land use practices, cause environmental degradation and other forms of resource depletion. This phenomenon is known as "ecological marginalization" (Homer-Dixon 1999). Since the dawn of time, societies have oscillated between damage to the environment and respect for the great laws of ecology.

The world population, estimated at nearly 2.6 billion in 1950, reached 7 billion in 2011 and will reach 9.7 billion by 2050. Since the industrial revolution in 1850, humans have acquired such an influence on the biosphere that humankind has become its central actor, to the point of changing our planet's appearance due to an excess of activity. This is conveyed by the notion of the Anthropocene, which etymologically means "the age of humans". Humans continue to overextend towards wildlife territories, which are unscrupulously being wiped out. As the natural functioning of ecosystems is disturbed, forests are cleared, soils are overturned and mined and species poached, we are increasing the risks of zoonoses and their evolution into epidemics or pandemics. In 2016, the United Nations Environment Programme (UNEP) noted that the increase in zoonotic disease epidemics worldwide was an increasing subject for concern. Studies particularly stressed that

75% out of all emerging infectious diseases in humans are zoonoses closely related to the health of ecosystems. This was clearly the case with the Covid-19 crisis.

2.3. The Mediterranean, more than a sea in the middle of the land

Gifted with a diversity of territories at the crossroads of three continents, its natural and cultural heritage and its climate, the Mediterranean region has experienced the severe impact of human development. This region has always been the scene of major environmental and development issues. The economy of this ecoregion is largely dependent on natural resources. Civilizations and dynasties have followed one another, thriving when environmental conditions were favorable and collapsing when resources were scarce.

Our ancestors who populated the Mediterranean basin were creators of biodiversity in the sense that they created new varieties by hybridizing and selecting plants. However, centuries of natural resource exploitation dramatically changed the appearance of the Mediterranean forever. This was demonstrated by a recent study in the south of France: even after two millennia of abandonment, the soil of the old Roman enclosures is today significantly rich in carbon and phosphorus, two elements provided by the droppings of sheep during their concentration in those enclosures (Saatkamp et al. 2020). As a result, even 2,000 years later, the vegetation in that area is different.

Practically all the ecosystems in the Mediterranean basin testify to the impact of recent human-induced changes in the environment. Major drivers of change include climate, population growth, pollution, unsustainable land and sea use practices and non-native species. In most regions, natural ecosystems and human livelihoods have been affected. As these factors continue to change globally and regionally, the impacts will be exacerbated in the coming decades. These "disturbances" shaking the symbiosis between humans and the biosphere have been accentuated by rural exodus in the hope of finding better quality of life and jobs in urban areas. While metropolises become increasingly populated, rural areas are becoming impoverished in labor and the land which was once cultivated is abandoned. Rural producers have turned into city dwellers with a consumer profile. It might seem that this shift in favor of urban areas is the turning point which led to the dismantling of the connection between the biosphere and humans. Even if they once established mastery over nature through an apprenticeship lasting for millennia, humans then became quickly distant from such learning.

The requirement to meet the needs of a growing population in a region whose climate is uncertain and where energetic crises are underway raises many issues:

availability and renewal of water resources, soil fertility, forest survival, natural balance and development of territories, etc. Globalization has also had a massive impact on biodiversity and its distribution due to the increased exchange of money, goods and services across borders, overlapping with various practices which ignore national and cultural borders. The (voluntary or accidental) introduction of exotic species by humans threatens ecosystems, habitats and native species, with negative consequences on ecology, the economy and health. Invasive species have appeared, and there is an increasing trivialization of fauna and flora.

The Millennium Ecosystems Assessment that the UN conducted in 2000 identified 17 categories of ecological services that biodiversity supplies to humans and warned about the seriousness of the stakes involved in maintaining the biodiversity of the planet's ecosystems. Political "awareness" about the need to preserve our biosphere has increased, while the MAB Programme invites researchers and actors to (re)teach humans how to cohabit with their environment. Theoretically, the declared goal is to reconcile sustainable development at the regional level with the protection of the environment and, more specifically, the conservation of biodiversity, with respect for the cultural values of all. There has been a retroaction involving the "reabsorption" of individuals who leave the urban environment due to pollution and move towards the rural environment in the search for green spaces. Biosphere reserves are thus considered privileged territorial spaces for experimenting with operating methods specifically adapted to the stated goals and cross-cutting issues.

In order to make humanity viable, it is essential to change our way of life. Nevertheless, these transformations have to be accompanied by demographic policies, for example, strengthening women's education and their autonomy, as well as family planning measures. The positive nuance is that advances in science and technology can greatly assist us in our quest to restore these altered links between humans and the biosphere.

2.4. The academic sphere and the action in favor of biodiversity

The academic sphere plays a pivotal role at this stage. The expansion of knowledge in the living sphere and the increase in threats to biodiversity (sources of this knowledge) imply that this area of research should be given priority. The new perspectives derived from scientific progress in the last quarter of a century concern the three main areas of exploration of biodiversity: the pursuit of the identification of new species, the progress of knowledge on genetic diversity (as well as functional and genome diversity) and determining the interrelationships between species within ecosystems.

Biosphere reserves could be living laboratories where this research can be initiated and optimized before being applied on a larger scale. This would involve a bottom-up approach; starting from invaluable local knowledge – which would otherwise be quickly lost – would encourage the development of appropriate management methods, taking into account the specificities of each living community (humans included). We can take advantage of the contributions of science and technology to sustainable development. Thanks to their diversity and their means of adaptation, not only do living beings provide us with chemical molecules to be used in pharmacopoeia, but they also offer biological solutions, far more sophisticated than human artifacts, in order to respond to selection pressures. This can significantly reduce the use and impact of the raw material and energy-intensive industrial processes currently at play.

According to the French Senate report on the contributions of science and technology to sustainable development, the "memory of 'success of the living' that biodiversity constitutes should lead to the rise of an industry based on biology and biotechnology which, together with the rise of nanotechnology, could be one of the mainsprings of the next industrial revolution. Two major scientific and technological projects will contribute to this: bio-inspiration and biotechnologies" (Laffite and Saunier 2007). It is in this sense that the academic field can help to address these challenges, by inviting an interdisciplinary approach between the different curricula. The knowledge accumulated by the social sciences can then be explored and studied by physicists, biologists or chemists and made concrete by engineers.

2.5. Biosphere reserves and Sustainable Development Goals

The 2030 Agenda for Sustainable Development and the Sustainable Development Goals (SDGs) associated with it were developed to succeed the Millennium Development Goals and were adopted in 2015 by the international community. With their design into three zones, biosphere reserves represent an excellent example where alternative human practices make it possible to establish a harmonious relationship between humans and nature, highlighting the contribution of natural resources and biodiversity to humanity. Biosphere reserves are considered as "natural gene banks" preserving the genetic diversity of plant and animal species, maintaining healthy ecosystems and natural landscapes. The importance of biosphere reserves could be emphasized by appreciating to what extent the majority of SDGs can be associated with this kind of protected area. The first three SDGs concern the eradication of poverty, the disappearance of hunger, food security, the improvement of nutrition and the promotion of sustainable agriculture and the promotion of good health and well-being and are directly related to the protection of natural habitats. These habitats contain the core natural resources that provide security to local communities against famine, scarcity or poverty and play an essential role in supporting livelihoods and well-being. Plant diversity, including the ancestors of cultivated plants and traditional varieties found in undisturbed areas such as biosphere reserves, provides local residents with the majority of their food needs, as well as a wide range of other crops and cultivated resources (Sharrock and Jackson 2017).

It is well known that many plants are used either directly as medicines or constitute the basis for medicines. As a matter of fact, it has been estimated that one plant in eight (about 45–50,000 species) has a medicinal use. According to the UN Comtrade Database[3] compiled in 2012, global exports of herbs with primarily pharmaceutical use were valued at USD 2.2 billion, whereas in 2000, the global sales of herbal products were valued at USD 60 billion.

Environmental education and the training in conservation, ecological restoration and sustainable development are key activities for many organizations involved in plant conservation. SDG 4, and specifically Goal 4.7, is particularly relevant for biosphere reserves. By 2030, the goal is to ensure that all students acquire the necessary knowledge and skills to promote sustainable development, in particular through education for sustainable development and lifestyles, human rights, gender equality, the promotion of a culture of peace and non-violence, global citizenship and appreciation of cultural diversity and the contributions of culture to sustainable development (Sharrock and Jackson 2017).

Section 6.5 in SDG 6 encourages the protection and restoration of water-related ecosystems, including mountains, forests, wetlands, rivers, aquifers and lakes. Biosphere reserves are privileged spaces where restoration techniques combining phytoremediation tools, new sanitation technologies and new waste management approaches can be tested before being applied on a large scale. The same is true for clean energy, which should benefit from the progress of our technologies and improve access to non-polluting and safe fuels for cooking, making it possible to increase the use of renewable energies other than electricity without abusing natural resources (Sharrock and Jackson 2017).

SDGs 8, 11, 12, 13, 14 and 15 all aim at reestablishing the sacred link between humans and nature and, in this case too, biosphere reserves are great open-air laboratories where a partnership with higher education institutions can only be beneficial for each of them. In this regard, the EduBioMed project was an incomparable opportunity that took the plunge and forged links between certain biosphere reserves and higher education institutions.

3 See: http://comtrade.un.org/db/.

2.6. References

Homer-Dixon, T. (1999). *Environment, Scarcity, and Violence*. Princeton University Press, Princeton.

Hublin, J.-J., Ben-Ncer, A., Bailey, S.E., Freidline, S.E., Neubauer, S., Skinner, M.M., Bergmann, I., Le Cabec, A., Benazzi, S., Harvati, K., Gunz, P. (2017). New fossils from Jebel Irhoud, Morocco and the pan-African origin of *Homo sapiens*. *Nature*, 546, 289–292.

Laffite, P. and Saunier, C. (2007). Les apports de la science et de la technologie au développement durable 2. La biodiversité : l'autre choc ? L'autre chance. Report, Office parlementaire d'évaluation des choix scientifiques et technologiques, 131.

Saatkamp, A., Henry, F., Dutoit, T. (2020). Romans shape today's vegetation and soils: Two millennia of land-use legacy dynamics in Mediterranean grass-lands. *Ecosystems*, 24, 1268–1280.

Sharrock, S. and Jackson, P.W. (2017). Plant conservation and the sustainable development goals: A policy paper prepared for the global partnership for plant conservation 1. *Annals of the Missouri Botanical Garden*, 102(2), 290–302.

3

Social Representations, Collective Organization and Mediterranean Biosphere Reserves

3.1. Introduction

This contribution presents a study of social representations by students engaged in the Erasmus+EduBioMed program, whose specific goal is to offer curriculum training for potential future biosphere reserve managers. This program aims to bring together various types of knowledge (scientists, managers, etc.) for this type of space and at the scale of the Mediterranean basin by targeting students who are culturally very different, with varied disciplinary backgrounds. The sharing of knowledge and know-how drawn from the experience acquired over time in various contexts by the different categories of the actors involved is considered key to improving governance and public participation processes in these areas (Beuret 2011).

Our hypothesis is that better preliminary shared knowledge regarding the social representations of the actors concerned – so as to reach a common vision of the future of a territory or a local development project, for example – promotes the proper functioning of knowledge dissemination mechanisms and the co-construction of projects implemented in the field. This is the claim of certain approaches in the social sciences, such as the patrimonial approach, for example, which studies the relationship between society and nature (Cormier-Salem 2014). Patrimonial

Chapter written by Angela BARTHES, Bruno ROMAGNY, Jean-Marc LANGE, Lahoucine AMZIL, Roser MANEJA, Mohammed ADERGHAL and Véronique CHALANDO.

approaches (de Montgolfier and Natali 1987; Babin et al. 2002; Petit and Romagny 2009, etc.) consider that the arising difficulties in natural resource management[1] mostly concern relationships between people and nature, and that they can be approached from the angle of coordination and collective action. For the supporters of the patrimonial approach, it is mainly the lack of communication that causes situations of environmental conflict. The discussion should make it possible to "put an end to a good part of the conflicts and [...] to favor a shift in the persisting ones" (Weber 1996).

Here, we introduce a first structural analysis of social representations regarding biosphere reserves. This analysis was conducted on the basis of a questionnaire submitted to master's degree students enrolled in different partner universities of the EduBioMed program in Morocco, Spain and France. These affiliations were set up when the program was established, but disciplinary coherence was not specifically addressed at this stage of the curriculum[2] construction. The students surveyed were invited from different master's courses: Geography at the University of Rabat, Ecology at the University of Barcelona and Earth and Life Science with a focus on teaching at the University of Montpellier. This multicultural and multidisciplinary panel of respondents enabled us to schematically identify similarities and remarkable differences between the social representations of these students, particularly in relation to differing national contexts.

This analysis adopts the perspective of critical scientific education, giving particular importance to a reflective view of the utilitarian ideology dominating our relationship with the world, in particular, when the initiatives emanate from European funding. In this view, Bader (2011) points out that critical theory "denounces the recourse to instrumental rationality based on a certain conception of

1 Resource management analysis is defined as "the joint study of an ecosystem and the interactions between actors in relation to this ecosystem, deliberately keeping a distance from an excessively technical conception. This definition presupposes, on the one hand, the existence of differentiated actors for whom the management of these resources is an issue. And on the other hand, it invites us to question the relationships between the dynamics of an ecosystem and the practices it supports, which are sometimes underpinned by conflicting representations and logics" (Aubert and Romagny 2012).

2 Curriculum is a Latin term reflecting the notion of a schooling career, and more broadly, all educational practices. This term was diffused by the American pedagogue Ralph W. Tyler in 1949 in his work *Principles of Curriculum and Instruction*. The word was then reabsorbed by the French language in Canada and made its appearance in France around 2014 with the reform of the school curricula. Source: https://fr.wiktionary.org/wiki/curriculaire.

science, framed within a positivist paradigm, privileging technology as a source for solutions [to environmental problems]" (Bader 2005). Roottom and Hart (1993) and Fortin-Débart and Girault (2009) remind us that this critical questioning of reality and this commitment to action are more effective when they are collective rather than individual. From this civic perspective, critical education would amount to an "education in power (Audigier 2005) aiming at providing skills to participate in the life of the city (public life, political life, daily life)" (Fortin-Débart and Girault 2009).

After having introduced some theoretical elements favoring a better understanding of the notion of social representations and the usefulness of this concept in a process of cross-curricular development, this chapter presents the methodological elements adopted to analyze the social representations of several groups of students on the subject of biosphere reserves. On the basis of the results obtained, in the last section, we propose some elements for discussion and debate on the role of this analysis of social representations within the framework of a European curriculum development program, such as EduBioMed.

3.2. Social representations as an exploratory method of prior knowledge

Knowledge of all kinds exists only because it is socially constituted, but it transcends the individuals or groups who produce it, use it and make it circulate. This differentiates it from knowledge attached to an individual or a group of individuals (Lange and Barthes 2017).

The social representation approach makes it possible to explore prior knowledge while acknowledging the role of the socio-cultural and historical context in the construction and sharing of representations (Moscovici 1961; Jodelet 1989; Abric 1994; Wagner 1994). We may argue that, despite there being an institutional and accepted definition of biosphere reserves, in practice, this object of study is complex and laden with values, subject to various interpretations. These must be clarified by the various actors who give them in order to succeed in fully investing in the forms of co-management generally proposed by biosphere reserves (Cibien 2006; Bouamrane 2007).

3.3. How can social representations be defined? Some theoretical elements

The act of representation is traditionally defined as "a presentation of something present in someone's mind" (Bonardi and Roussiau 1999) and is considered a real way of knowing reality. It is a process that involves the internalization of an object by a subject, or a group, thereby becoming an object of thought whose content replaces reality. However, this reconstruction by the subject is inseparable from individual and collective history, from the social and ideological context. In this sense, representation is always social insofar as the act of representation makes it possible to integrate both the objective characteristics of the object represented, the specificities of the subject or group, and the elements from its normative context. Social representations constitute a powerful tool for analyzing the normative systems of a group at a given time.

Social representations can also be defined as a "functional vision of the world" (Abric 1994) in the sense that, by providing the individual or group with an intelligible reading of the complex world, they make it possible to position ourselves in relation to the normative system and to build our own frame of reference. This positioning is part of three interdependent processes: (I) interpretation and therefore reconstruction of reality, (II) communication with the group and (III) mastering a code of conduct in the environment.

Social representations grant the communication function between individuals in the sense that they make it possible to develop a common code – herein called the "normative framework" – to name and classify their knowledge and interpretations of the world in the most unequivocal way possible. This normative framework is also a guide for action insofar as, through communication, it makes it possible to interpret daily reality, to decide on its various aspects, to position ourselves and, if necessary, to take a decision in relation to these aspects (Jodelet 1989).

Therefore, the practical goal of social representations and the control of the environment become real challenges. Social representations are defined as "a form of socially developed and shared knowledge – common sense knowledge – having a practical goal and contributing to the construction of a reality common to a social group" (Jodelet 1989). This form of knowledge, frequently called "natural knowledge" in Education Sciences (Barthes and Alpe 2016), is directly linked to the behavioral forms adopted by the group or the individual (Barthes et al. 2015).

3.4. How can social representations be defined? Central core and peripheral elements

Studying a social representation and understanding how it works requires exploring its content and structure. According to Abric's (1994) Central Core Theory, any representation is organized around a central core, without which there can be no real representation and which gives the representation its meaning and coherence. The central core is structuring, assumes an obvious status for the group or the individual, and constitutes a stable element. The notion of core also refers to the identification of individuals with a social group, considering that it guarantees its homogeneity. The core is determined not only by the nature of the object represented but also by the relationship the subject or the group maintains with it and, finally, by the system of values and norms defining the ideological context (Abric 1994). It is therefore the identification of the central core that enables the comparative study of representations. The peripheral elements are organized around this core. These make it possible to turn the new information resulting from a given and contextualized situation into something concrete, understandable and transferable.

The peripheral elements are organized around the central core. When they are close to the central core, they play an essential role in concretizing the meaning of representation. Distant from the core, they illustrate or justify this meaning. Peripheral elements constitute the most accessible set of social representations because they are less stable and more alive. In concrete terms, the new information is mostly elaborated by the peripheral elements. They constitute a "set of pronounced judgments made about the object represented and its environment, stereotypes, beliefs" (Abric 1994). They have an essential role, and this role will be broadly evidenced in the link between social representations and education in sustainable development because they play a role "of an interface between the central core and the concrete situation in which the representation is elaborated or operates" (Abric 1994).

The peripheral elements are directly dependent on the context and result from the anchoring of representation in reality. Given the stability of the central core, they constitute the moving and evolving dimension of the representation, while ensuring the defense of the central core. This defense function has been called the "bumper" by Claude Flament (1987), insofar as it is through the modification of its periphery that the transformation of a social representation can take place, possibly at a later stage. Flament considers that the peripheral elements are "schemes" organized by the central core, "instantly ensuring the functioning of representation as a grid for deciphering a situation". They prescribe behavior and act as an instantaneous guide on how to react to a specific situation, by designating what is normal (to do or say) in a particular situation. In this way, they also guide the reactions or the actions of the subjects instantaneously, without the recourse to core meanings. They constitute a personalized modulation of the central cores and are socially developed and shared in an ampler manner.

Social representations are therefore a dual system, comprising a central core and peripheral elements: the central core is bound to the historical, sociological and ideological context. It is strongly marked by the group's collective memory. It defines the group's normative framework. Associated with values and norms, it defines the fundamental elements around which the consensual representations relatively independent from immediate contexts are built. Thus, "the core system is stable, coherent, consensual and historically marked" (Abric 1994). The peripheral system is more individualized and contextualized, that is, more associated with individual characteristics and the immediate environment. Unlike the essentially normative core system, the peripheral system is, for its part, mainly functional. It is flexible and can enrich the representation with individual variations derived from the subjects' own stories, personal experiences or lived events. This function truly guarantees the possibility of developing individualized social representations, nonetheless organized around a common central core. "The peripheral system is the essential complement to the core system on which it depends. The functioning of the core can only be understood in continual dialectic with the periphery" (Flament 1994).

3.5. The methodological elements of our study

To carry out the present study of social representations in relation to biosphere reserves, we chose the quantitative approach implemented by Pierre Vergès (1994, 2001) due to its reproducible nature, which makes comparisons possible. The data presented here are the answers provided to a spontaneous evocation questionnaire administered to three groups of master's students in France, Spain and Morocco. The surveyed population included 41 students enrolled in Morocco, 38 students in France and 39 students in Spain, making a total of 118 students aged between 23 and 27. This sample is not representative in the statistical sense, although it makes it possible to identify certain trends at the level of the population studied. The main question[3] posed to the students was the following: "Which words[4] or expressions spontaneously come to your mind when you think about biosphere reserves?"

3 Although we were interested in the evocations aroused by other notions, such as patrimony or biodiversity, for example, in this text, we chose to focus on the notion of the biosphere reserve because it is at the heart of the concerns of the EduBioMed Program.

4 Respondents had to provide a minimum of five words or expressions and a maximum of ten.

Two preliminary stages were necessary to analyze the answers given by the students to this question. We first categorized the elements evoked by the subjects in function of their lexical proximity. At a later stage, two indicators were taken into account: the frequency of appearance of each of the terms in the responses of the surveyed population, and the co-occurrences (number of links between terms). It is the crossing of these two indicators which enables us to offer a graphical representation. The organizational graph (Barthes and Alpe 2016) makes it possible to better grasp the notion of knowledge organization and to modify the aforementioned social representations at will. This operation involves constructing a notional network in graphical form. This implies establishing relationships between the different items by analyzing similarities (Degenne and Vergès 1973; Vergès and Flament 1997). The more traditional approach involves the spontaneous classification of items by the respondents.

Furthermore, the issue of recoding terms, that is, grouping them together, is a methodological point that requires special attention. Recoding is often necessary, if only initially to group similar meanings among the answers provided. The most reliable method is direct analysis, which involves constructing co-occurrence tables from spontaneous responses, without recoding. However, this technique often proves impossible to implement because the responses obtained can be very similar. The method we used is controversial in our field of research and is not unanimously shared.

For this reason, we had to keep a critical eye on what we did at this stage in order not to forget the different recodings performed (which can end up creating several types of recodings, depending on the study's goals) and to bear in mind the limitations of the different techniques. This graphic approach is based on the hypothesis that the sum of the co-occurrences over a sufficient number of individuals shows the permanence of significant links in the object of representation.

We should bear in mind that it is possible to use manual, automatic or semi-automatic methods to conduct this type of analysis. The manual methods involve placing all the words along the abscissa and ordinate of a table, then noting the number of co-occurrences at the intersection of the words. For example, if individual one mentioned the terms X, Y and Z in a spontaneous evocation question; individual two the terms X, Y and W; and individual three the terms X, Y, Z and U, then the response table will be as given in Table 3.1.

	X	Y	Z	W	U
X	0	3	2	1	1
Y	3	0	2	1	1
Z	2	2	0	0	1
W	1	1	0	0	0
U	1	1	1	0	0

Table 3.1. *Example of the construction of a co-occurrence table*

Table 3.1 reads as follows: there are three co-occurrences of the terms X and Y, that is, three respondents who answered X also answered Y. This comes down to stating that there are no co-occurrences of the terms W and U, that is, none of the respondents who answered U answered W, as well.

While graphic processing can be done manually, most of the classic software on the market enables automatic data processing. In our case, we used Modalisa software (Kynos).

The graphical results are presented as networks of words linked together. They represent an image of the social representation's notional network. Strongly interconnected and frequently quoted words were presumed to be part of the central core; the closer and more weighted the item connections, the closer and more important is the periphery in the interpretation of social representations.

To comment on a graph of social representations, it is first necessary to define the central core and the various peripheries before explaining their signifiers. The organizational graph provides information not only regarding link proximity between the items but also between the items and the central core and its peripheries.

3.6. Study results

Figures 3.1–3.3 summarize the main results of our analysis for each group of students surveyed.

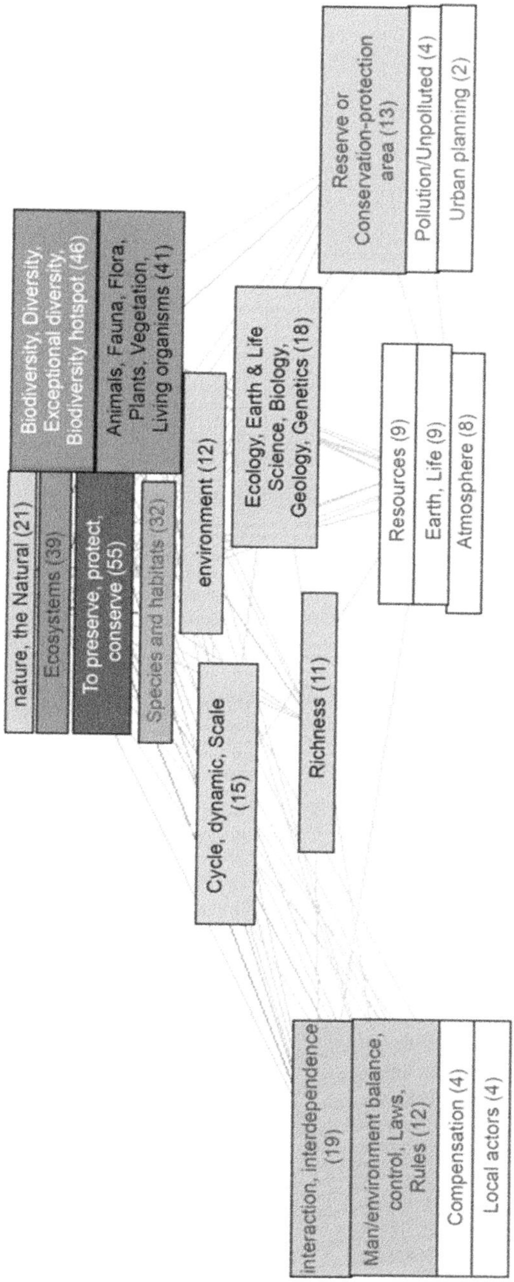

Figure 3.1. *The evocations of the students from the teacher training, Earth and Life Science master's degree (France)*

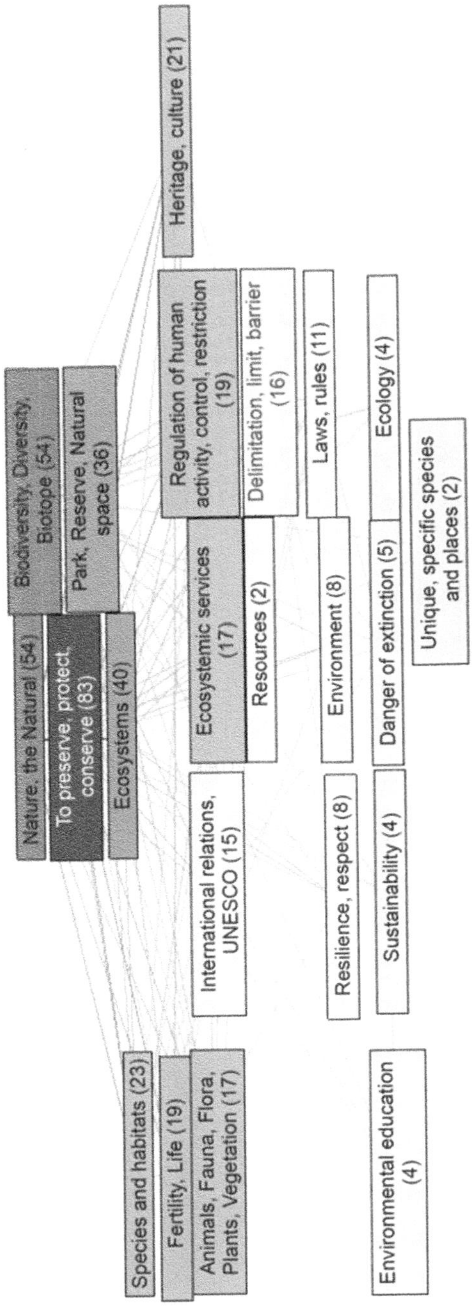

Figure 3.2. *The evocations of the students from the Ecology master's degree (Spain)*

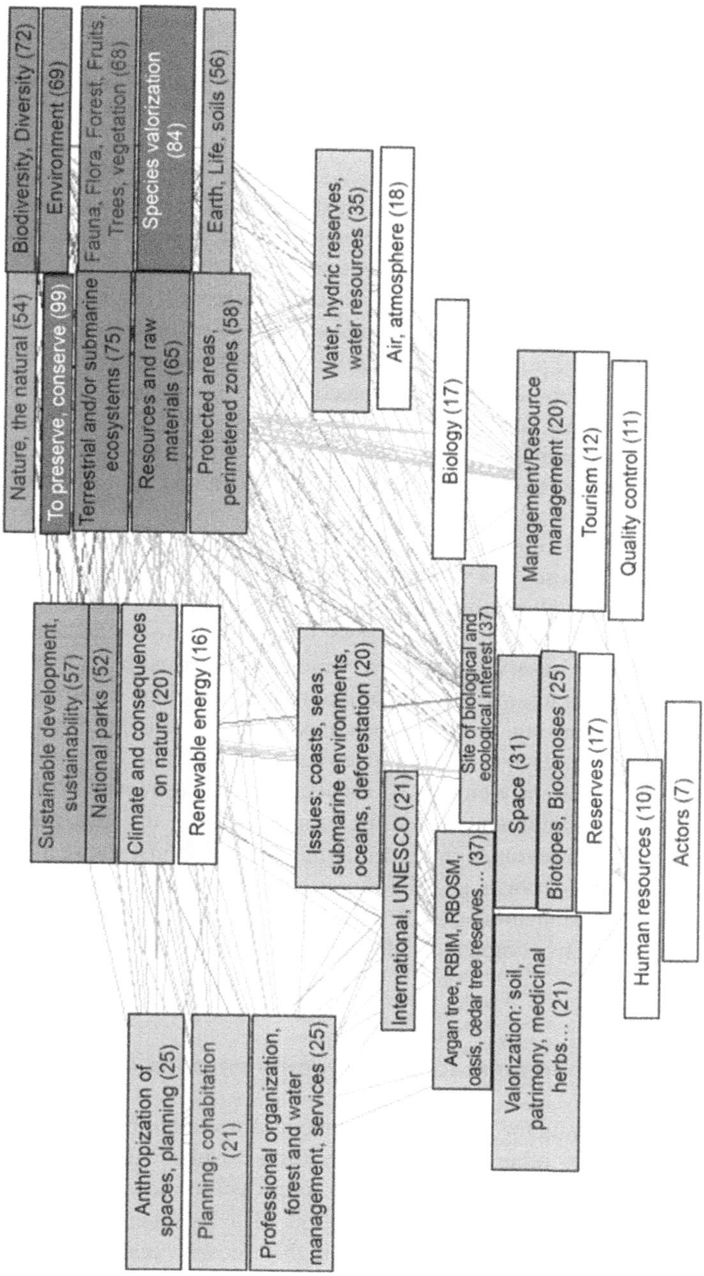

Figure 3.3. *The evocations of the students from the Geography master's degree (Morocco)*

3.7. Differences and similarities in the social representations of students

3.7.1. *Social representations relatively shared by students enrolled in France and in Spain*

From a comparative perspective, the analysis of the corpus differences and similarities between the three national contexts showed a relative agreement between the master's degree students in Spain and in France. The responses provided by these two groups suggest the hypothesis of an existing relatively shared social representation, especially considering that the reference disciplines are not distant from one another. Social representations seemed to be essentially structured around nature as the anchor point for the notion of the biosphere reserve, associated with those of preservation/conservation/protection of biodiversity, species and habitats. The only difference observed at the level of the central core concerned the category "Fauna, flora, animals, species and habitats", which was present among students enrolled in France but appears more peripherally among Spanish students.

This notion of nature, in connection with the biosphere and its protection, seems to constitute the central core of a possible representation. The peripheries were a little different, more factual among students enrolled in France, more functional among those enrolled in Spain. In the first case, relatively close to the central core appeared the notions of reserves or conservation areas, wealth and resources, human/environment interactions with variations concerning controls, laws, regulations, compensation, etc. In the evocations made by the French group of students, we observe the presence of disciplines such as Earth and Life Science and Ecology, which are important elements to associate with the evocation of cycles, dynamics and spatio-temporal scales. For the group in Spain, the near peripheries were more functional and revolved around international relations, the regulation of human action, ecosystem services and the rhetoric of sustainable development (the danger of extinction, sustainability, etc.). We can note an interesting mention of heritage and culture. Environmental education, on the other hand, is rarely mentioned.

3.7.2. *Notably different results for the students enrolled in Morocco*

The evocations' analysis for the group of students enrolled in Morocco provided different results from the other students. First, these results were richer and the evoked words more interconnected. The central core was more complete, showing a more systemic vision of the social representation of a biosphere reserve. In the central core, the anchoring of nature was also highly present, just as the notion of biodiversity was strongly associated with its preservation/conservation. However, other elements constitute this central core. In this context, resources and raw materials, the valuation of species, the environment, ecosystems and the Earth, life and water systems were additional elements which completed the central core for this group of students.

The issue of water and water reserves was widely mentioned at the level of the near periphery, showing that biosphere reserves are associated with challenges to the natural environment (coasts, seas, deforestation, etc.), their perceived anthropization and urban planning. The practical issue of professional organization that gravitates around biosphere reserves also stood out in relation to planning and cohabitation issues.

However, the most salient fact for this group of students seemed to be the apprehension of the question of the biosphere reserve at a larger scale, within a logic that could be characterized as systemic. Note that, in this case, the reference discipline was Geography, which could also account for this result. In fact, the question of the scales was openly stated, with a special mention of the international level (UNESCO, among others) and the matching national systems (argan tree reserve, oasis, etc.). In general, the students enrolled in France seemed to have more difficulty quoting concrete examples of biosphere reserves at the national level compared to those enrolled in Morocco, who, on the contrary, could do it quite systematically and in a relatively exhaustive way.

A final highlight for this group of students was that they explicitly mentioned the direct applications to sustainable development and ecological transition issues (climate, food, etc.). Biosphere reserves are well understood in this global environment, directly linked to the challenges of promoting local products and the (material and immaterial) patrimony through alternative tourist activities (rural tourism, ecotourism, etc.).

3.8. Addressing the issue of complexity versus focusing on the environment

The analysis of the results for this study on the social representations of biosphere reserves raises several tensions.

The first tension concerns the difference in apprehension of the level of complexity of the issues surrounding biosphere reserves by the three groups of students surveyed. This complexity was highly present among students enrolled in Morocco, supposed to be the recipients of the curricular elaborations (Forquin 2008) piloted by the French and Spanish teams within the framework of the EduBioMed program. Within the context of the EduBioMed program, the Spanish and the French have to offer a training program to the Moroccans. On the other hand, the students enrolled in France and in Spain seemed to have a rather simplified preliminary vision of this notion of complexity.

Addressing the collective organization of society and the systemic and sustainability approaches around biosphere reserves would remove the obstacle to critical reasoning, particularly if this emanates from those responsible for developing the curricula. Our hypothesis is that the trend is to bring up and then to minimize the question of the domination of the values disseminated by Europe in the recipient countries. The social representation of biosphere reserves from the three corpora studied presents strong focal points and corroborates the results previously obtained in France regarding sustainable development issues (Barthes 2011). The focus on nature and biodiversity, particularly among the students enrolled in Spain and in France, is likely to limit their understanding of the complexity of the processes at play around biosphere reserves, which also have a political, cultural and social gradient. Given the fact that these social representations strongly focus on nature, the risk of apprehending the issue of biosphere reserves without recognizing all its complexity together with its multiple socio-ecological dimensions is overwhelming.

The analysis of the focal points of social representations, in addition to the remarkable absence of political, economic and institutional issues from the central core of the representation (more particularly among the students enrolled in France and in Spain), confirmed its fragmented character, insufficient to apprehend the complexity of the concepts taught. Although some economic aspects were mentioned by students in France and in Morocco, these only appeared among the peripheral elements: in the first case, in relation to the limited character of resources and in the second case to their their valorization. The notion of compensation was mentioned among the students enrolled in Spain. These elements were not organizing elements of representation, despite being central components of biosphere reserves. While students in Spain and France took little account of the role of political and institutional bodies in the promotion of sustainable development, the students in Morocco seemed to be more aware of this dimension, especially when it comes to international cooperation elements.

Finally, in relation to the analysis, none of the groups of students envisaged adopting any form of politico-citizen action (neither in their social life, nor in their future professional life).

At this point, we would like to emphasize that the absence of a debate on the potentially controversial nature of the concept of the biosphere reserve was patent. The representations of the three groups of students seemed to show that they overlooked the critical dimensions, the alternative models, the complexity elements which are so characteristic of biosphere reserves. In addition, the focus on nature was centered on a truncated representation of the biosphere reserve, especially among the students enrolled in Spain and in France. Furthermore, the students' comments in terms of resources and valorization methods confirmed a vision of sustainable development policies centered on a reductive vision of the environment, exclusively understanding resources as something at the service of human beings. The critical approach to education should call into question this utilitarian vision. Otherwise, we risk using the educational discourse as an ideological tool to back up a certain vision of the world.

3.9. Addressing the collective organization of society versus the recourse to individual action

Still, there was a difference between the corpus of students enrolled in Morocco (whose social representations constituted a less marked and more nuanced obstacle to critical education) and those of the students enrolled in France and in Spain. As a matter of fact, the focusing elements did not contemplate the same dimensions.

This behavioral focus element constituted an additional hurdle, making it practically impossible to keep a critical distance towards the real. The question then arises as to whether – even at university – teaching is not sometimes a "moral course" favoring political correctness to the detriment of knowledge (Legardez 2006).

Indeed, the prominence of individual action and responsibility, present in the second periphery, overlooks the problem of the collective and the systemic organization of society. In fact, it raises the question of the non-distancing of education with regard to proximity social practices (Alpe 2006) and ends up representing an obstacle to the critical perspective in the teaching of sustainable development perspectives. In short, a critical approach to education requires working beyond the level of local and individual action. This type of obstacle-focus was not present among the students enrolled in Morocco. Their social representations were more oriented towards the collective organization of society and less towards a problem that should be dealt with.

3.10. Conclusion

In light of the proposed approach and our results, two key conclusions can be drawn. First, we can hypothesize that, depending on the socio-cultural and political context, on the one hand, and on the disciplinary context, on the other, the biosphere reserve invites strikingly different social representations. In terms of corpus deviations, we were able to identify a focus on the "natural" aspects among the students in Spain and in France, and an emphasis on the social aspects and the international cooperation dimension among the students in Morocco. However, in the three national corpora, the social representation of biosphere reserves was strongly focused and had a highly fragmented character. These characteristics impede a global vision of the object, as well as of the systematization of the elements of representation and therefore of its complexity. The representational elements constitute important socio-educational specificities that should be taken into account when we wish to engage university training in biosphere reserves. In addition to this observation, the societal challenges are major and relate to the European Union's presentation of sustainable development as one of its fundamental and transversal goals.

In order to avoid the risk of merely training managers at the service of governing bodies, whatever these may be, and in accordance with the goal of "training critical citizens", teaching should aim at enriching the students' representation. This can be achieved by integrating complex elements, by acknowledging their mutual interrelationships and the issues that arise from them, in what could be described as a new "system of representations-knowledge" (Legardez 2004). We can indeed consider that the role of education in general (and of higher education institutions in particular) is not only to provide a certain threshold of employability (Adomssent 2006) to students but should go beyond this need and contribute to forming both scholarly and critical citizens.

3.11. References

Abric, J.C. (1994). *Pratiques sociales et représentations*. PUF, Paris.

Adomssent, M. (2006). *Higher Education for Sustainability. New Challenges from a Global Perspective*. Springer, Berlin.

Alpe, Y. (2006). Quelle est la légitimité des savoirs scolaires ? In *L'école à l'épreuve de l'actualité, enseigner les questions socialement vives*, Legardez, A. and Simmoneaux, L. (eds). ESF, Paris.

Aubert, P.M. and Romagny, B. (2012). De l'économie néo-institutionnelle et patrimoniale à la sociologie de l'action organisée. In *Agdal. Patrimoine socioécologique de l'Atlas marocain*, Auclair, L. and Alifriqui, M. (ed.). IRD-IRCAM, Rabat.

Audigier, F. (2005). Les enseignements d'histoire et de géographie aux prises avec la forme scolaire. In *Les formes de l'éducation : variété et variations*, Maulini, O. and Montandon, C. (eds). De Boeck Supérieur, Brussels.

Babin, D., Antona, M., Bertrand, A., Weber, J. (2002). Gérer à plusieurs des ressources renouvelables. Subsidiarité et médiation patrimoniale par récurrence. In *Patrimonialiser la nature tropicale. Dynamiques locales, enjeux internationaux*, Cormier-Salem, M.C., Juhe-Beaulaton, D., Boutrais, J., Roussel, B. (eds). L'Harmattan, Paris.

Bader, B. (2005). Rapprochement interdisciplinaire entre éducation aux sciences, citoyenne et l'éducation relative à l'environnement : points de vue de chercheurs et formation des enseignants. In *Éducation et environnement. Un croisement de savoirs*, Sauvé, L., Orellana, L., Van Steenbergue, É. (eds). Fides, Montreal.

Bader, B. (2011). Éducation à l'environnement dans une société du risque : la conception des sciences privilégiée de l'éducation au développement durable. In *Éducation, environnement et développement durable : vers une écocitoyenneté critique*, Bader, B. and Sauvé, L. (eds). PUQ, Quebec.

Barthes, A. (2011). Comment réintroduire les savoirs face à l'éducation au développement durable ? Exemple des filières professionnelles d'aménagement des territoires. *Actes du colloque. Éducation au développement durable et à la biodiversité : concepts, questions vives, outils et pratiques*. Digne-les-Bains, 20–22 October [Online]. Available at: http://www.refere.uqam.ca/FR/publications_mono.php.

Barthes, A. and Alpe, Y. (2016). *Utiliser les représentations sociales en éducation*. L'Harmattan, Paris.

Barthes, A., Jesiorski, A., Legardez, A. (2015). Les études européennes : un cadre favorable à la mise en œuvre de l'éducation au développement durable dans une perspective critique ? In *Éducation au développement durable. Enjeux et controverses*, Diemer, A. and Marquat, C. (eds). De Boeck Supérieur, Brussels.

Beuret, J.E. (2011). Quelles voies pour la participation du public à la conservation de la nature ? Le cas des réserves de biosphère. *Développement durable et territoires*, 2, 3 [Online]. Available at: http://journals.openedition.org/developpementdurable/9096.

Bonardi, C. and Roussiau, N. (1999). *Les représentations sociales*. Dunod, Paris.

Bouamrane, M. (ed.) (2007). Le dialogue dans les réserves de biosphère : repères, pratiques et expériences. Report, UNESCO, Paris.

Cibien, C. (2006). Actualités de la recherche. Les réserves de biosphère : des lieux de collaboration entre chercheurs et gestionnaires en faveur de la biodiversité. *Natures Sciences Sociétés*, 14, 84–90.

Cormier-Salem, M.C. (2014). Représentations sociales de la biodiversité et implications pour la gestion et la conservation. In *Sciences de la conservation*, Gauthier-Clerc, M., Mesléard, F., Blondel, J. (eds). De Boeck Supérieur, Brussels.

Degenne, A. and Verges, P. (1973). Introduction à l'analyse de similitude. *Revue française de sociologie*, 14, 471–512.

Flament, C. (1987). Pratiques et représentations sociales. In *Perspectives cognitives et conduites sociales 1. Théories implicites et conflits cognitifs*, Beauvois, J.L., Joule, R.V., Monteil, J.M. (eds). PUF, Paris.

Flament, C. (1994). Structure, dynamique et transformation des représentations sociales. In *Pratiques sociales et représentations*, Abric, J.C. (ed.). PUF, Paris.

Flament, C. (1995). Approche expérimentale de type psychologique dans l'étude d'une représentation sociale. *Les cahiers internationaux de psychologie sociale*, 28(4), 67–76.

Forquin, J.C. (2008). *Sociologie du curriculum*. Presses universitaires de Rennes, Rennes.

Fortin-Debart, C. and Girault, Y. (2009). De l'analyse des pratiques de participation citoyenne à des propositions pour une éducation à l'environnement. *Éducation relative à l'environnement : regards-recherches-réflexions*, 8, 129–145.

Jesiorski, A. (2010). Représentations sociales du développement durable d'étudiants en master études européens. Étude comparative : France, Allemagne, Pologne. Master's thesis, Aix-Marseille University, Marseille.

Jodelet, D. (ed.) (1989). *Les représentations sociales*. PUF, Paris.

Lange, J.M. and Barthes, A. (2017). Savoirs. In *Dictionnaire critique des enjeux et concepts des "éducations à"*, Barthes, A., Lange, J.M., Tutiaux-Guillon, N. (eds). L'Harmattan, Paris.

Legardez, A. (2004). L'utilisation de l'analyse des représentations sociales dans une perspective didactique. L'exemple des questions économiques et sociales. *Revue des sciences de l'éducation*, 30(3), 647–665.

Legardez, A. (2006). Enseigner les questions socialement vives, quelques points de repères ? In *L'école à l'épreuve de l'actualité, enseigner les questions socialement vives*, Legardez, A. and Simmoneaux, L. (eds). ESF, Paris.

de Montgolfier, J. and Natali, J.M. (eds) (1987). *Le patrimoine du futur. Approche pour une gestion patrimoniale des ressources naturelles*. Economica, Paris.

Moscovici, S. (1961). *La psychanalyse, son image et son public*. PUF, Paris.

Petit, O. and Romagny, B. (2009). La reconnaissance de l'eau comme patrimoine commun : quels enjeux pour l'analyse économique ? La mise en patrimoine de l'eau. *Mondes en développement*, 37(145), 29–54.

Roottom, I. and Hart, P. (1993). *Research in Environmental Education. Engaging the Debate*. Deakin University Press, Geelong.

Vergès, P. (1994). Approche du noyau central : propriétés quantitatives et structurales. In *Structures et transformations des représentations sociales*, Guimelli, C. (ed.). Delachaux et Niestlé, Neuchâtel.

Vergès, P. (2001). L'analyse des représentations sociales par questionnaires. *Revue française de sociologie*, 42(3), 537–561.

Vergès, P. and Flament, C. (1997). Comparaisons de méthode : à propos de l'article "Apport des modèles graphiques gaussiens en analyse de similitude". *Papers on Social Representations*, 6(1), 73–87.

Wagner, W. (1994). Fields of research and socio-genesis of social representations: A discussion of criteria and diagnostics. *Social Science Information*, 33(2), 199–228.

Weber, J. (1996). Conservation, développement et coordination : peut-on gérer biologiquement le social ? *Communication au colloque : "Gestion communautaire des ressources naturelles renouvelables et développement durable"*. Harare, 24–27 June.

4

Challenges and Opportunities of Collaborative Research on Biosphere Reserves in the Mediterranean

4.1. Introduction

The Mediterranean basin, a dynamic region with asymmetries between its northern and southern shores, diverse economic, political and cultural systems, and numerous sub-regional conflicts, includes 136 biosphere reserves (Escribano 2000; Olson and Olson 2013). Collaborative research across the Mediterranean focusing on these biosphere reserves is hampered by a lack of a common language and working styles that do not seem to be alleviated by increasingly sophisticated cyber infrastructure (Stuart et al. 2018). Furthermore, researchers in the Mediterranean face problems common in many parts of the world, such as the lack of quick and simple tools for efficient data sharing, as well as limited information and resources. Even globally accessible tools require time and information, which are not readily available to researchers attempting to store, publish and share data (Stuart et al. 2018). The aim of this chapter is to present global challenges and regional particularities of the Mediterranean that may impede collaborative research on biosphere reserves in the region.

4.2. Collaborative research

Researchers have formed and joined collaborative research groups and communities to address complex issues and to mitigate knowledge gaps by sharing

---

Chapter written by Moustapha ITANI, Salma NASHABE TALHOUK, Wassim EL-HAJJ, Nivine NASRALLAH and Hannah ABOU FAKHER.

competencies and resources, pooling expertise, participating in knowledge co-production and engaging in multidisciplinary research, which allows for more transparency and less secrecy (Meleis 2000; Shepherd et al. 2016; Rycroft-Malone et al. 2016). The nature and scope of collaborative research may be assessed through qualitative or quantitative approaches that analyze indicators such as the co-authors' professions, institutional affiliations, the organizational level of the collaboration, the authors' disciplinary focus and/or the authors' geographical focus (Amabile et al. 2001; Sonnenwald 2007; Hu and Racherla 2008). Although not all fields lend themselves to collaborative research, the number of peer-reviewed publications with multiple co-authors is on the rise, with twice as many publications reported in 2015 (21.3 %) than in 2000 (10.7 %) (Qin et al. 1997; Ribeiro et al. 2017; Koporc 2018; Øye et al. 2019). The number of countries participating in these collaborative networks also reportedly increased from 174 in 2000 to 200 in 2015, although the level of collaboration varies from one country to another (Leydesdorff and Wagner 2008).

4.3. Beneficial aspects of collaborative research

Researchers have worked together in collaborative groups for tens of years, and this trend has become mainstream, with increased research complexity and costs to deliver unique knowledge and skills not otherwise available but essential to research outcomes (Beaver and Rosen 1978, 1979a; Thorsteinsdottir 2000). Researchers pool resources to jointly purchase and use costly equipment, without which research would be less efficient or impossible, and funding agencies are driving collaborative research by facilitating collaborations as part of their funding conditions (Meadows 1974; Beaver and Rosen 1979b; Thorsteinsdottir 2000). An increase in the number of individuals participating in research is another benefit, as collaborative projects draw in people who have previously not been involved and who might not have contemplated initiating a research program. By extending the circle of researchers, perspectives broaden, and more expertise is added to the field. Besides the superordinate goal of the collaboration, collaborative research has a social context where partners fulfill their individual goals (Sonnenwald 2007). In many cases, collaboration is the key mechanism for mentoring graduate students and postdoctoral researchers and enhancing the productivity of individual scientists (Melin 2000; Bozeman and Corley 2004).

Publishing productivity may not be directly related to collaborative research, however, in the presence of moderating variables; the simple number ("normal count") of peer-reviewed journal papers is strongly and significantly associated with the number of collaborators (Lee and Bozeman 2005). It is important to note that many studies simply could not be carried out by individual scientists working on their own, especially not in a research environment in which science is increasingly interdisciplinary, equipment-dependent and project-based.

Data sharing offers a wide range of advantages for people engaged in scientific inquiries and science-based applications, such as nature conservation, whether professionals, hobbyists or researchers. Data sharing enables data to be added to existing datasets, meta-analyses, validation and testing of the reproducibility of studies and policy and strategy development. When data sharing occurs through cloud platforms, the information becomes accessible to practitioners. Although there are constraints that arise with transparency which relate to confidentiality, the differing interests of collaborators and informed consent, data sharing remains a beneficial aspect of collaborative research (Anderson 2013; Nielsen 2013; Foe and Larson 2016).

4.4. Challenges to collaborative research and data sharing

Impediments to collaborative research include geographic distance, which entails significant travel costs and time for researchers to discuss, in person, conceptual aspects of the research (Goddard et al. 2006). The distance resulting in different time zones further complicates communication, favoring emails rather than face-to-face or telephone communication. Dependence on funding and the lack of it affect researchers' abilities to network at international conferences and workshops (Morrison et al. 2003).

Research contexts (countries and cultures) may influence consistency in project focus and deliverables, especially when partner institutions have different requirements and processes. As a result, co-investigators end up receiving different types of information and following different approval procedures. Lack of training, resulting in poor data organization skills (Stuart et al. 2018) and limited knowledge about data sharing and management, gives room for misinterpretation and misuse of data (Tenopir et al. 2011; Kervin et al. 2014). Limited expertise in information technology (IT) and the "Internet of Things" limits the possibilities for scientists and practitioners to share data and collaborate with others simply because they do not know where and how to share data (Stuart et al. 2018). Legal constraints differ between the countries[1] of collaborating partners concerning copyright, intellectual

1 To find the appropriate database in a universe of cyber infrastructure, the Registry of Research Data Repositories has committed their work to facilitating the availability of data for scientists (https://re3data.org). This Web platform provides over 1,000 data repositories from multiple domains (Pampel et al. 2013, 2014). It consists of a search engine that acts as a gateway to many registered databases, such as the Northern Ontario Plant Database, for example, or the India Water Platform, which are two databases from different domains (flora and water management).

property rights and confidentiality (Uhlir and Schröder 2007; Michener 2015). The absence of long-term funding and support for the management of data and online platform maintenance may lead to their discontinuation.

4.5. Motives behind collaborative research

Several reasons for the researcher to engage in collaborative research were defined conceptually by Beaver and Rosen (1978), and these include access to special equipment and facilities, access to special skills, access to unique materials, access to visibility, efficiency in the use of time, efficiency in the use of labor, to gain experience, to train researchers, to increase productivity, to multiply proficiencies, to avoid competition, to surmount intellectual isolation, the need for additional confirmation of the evaluation of a problem, the need for stimulation of cross-fertilization, spatial propinquity, and through accident or serendipity. More recently, a survey with university professors revealed that motives behind collaborative research were primarily pragmatic, such as the special competence of the collaborator (41%), special data/equipment they owned and the development and testing of new methods (9%) (Melin 2000). The author also reported social motives behind collaborative research, such as friendships and successful past collaborations (16%), as well as relations with students' supervisors (14%). Regarding the benefits of collaborative research, these revolved around enhanced knowledge (38%) and research quality (30%), networking (25%) and the generation of new ideas (17%).

4.5.1. *Components of collaborative research*

Components leading to effective collaborative research were identified by Wine et al. (2017) following a review of collaborative research on environmental health. The authors identified external or institutional components that impact prospects of collaboration and interdisciplinary work and internal components inherent to the operational dynamics within a team and related to team productivity. Below is a summary of external and internal best practices relayed by Wine et al. (2017).

4.5.2. *External components*

Adequate funding to support interdisciplinary participation is essential because collaborative research is a lengthy process that requires traveling, meeting events and appropriate technology to allow for communication, data collation, licensing and network sharing. Acknowledgments and reward systems for participants engaged in data sharing should be based on clear guidelines and supported by standardized cyber policy, including for copyright and intellectual property, which differ between countries and projects.

Training of participants in new tools of collaborative research is important and should be planned and accounted for during the formation of a collaborative research team because research tools are continuously being updated and new ones introduced.

Institutional support for including communities in collaborative research by hiring community members to participate in collaborative research projects is crucial and requires institutional endorsement at the level of the ethics board, and policies need to favor such practices and promote data sharing with participants to maximize transparency and efficiency.

4.5.3. *Internal components*

Allocating time and the equitable distribution of resources for networking through a fair distribution of funds are necessary to maintain harmony in the group and allow bonding between collaborators, which is a long-term process that requires travel and time commitment beyond the assigned project meetings.

Addressing disciplinary and sectoral issues can often be challenging when certain disciplines dominate over others in dictating approaches; when collaborators from different institutions and regions, and sometimes with different languages, have divergent ideas, goals and priorities; or when discipline legitimacy, credibility and scientific independence are priorities for individuals in the group. This could inhibit the effectiveness of team efforts if proper cooperation is not in place to ensure that collaborators remain motivated to work together towards one common goal.

Building relationships is a pillar of the entire collaborative research process and ensures that trust and professional relationships maximize motivation for knowledge co-production and help address power imbalances.

Ensuring representation of all stakeholders to maximize the potential of collaborative research can be a challenge due to either poor stakeholder representation or the refusal of some to collaborate.

Embedding participation in the research is essential to enhance participation in research through the early involvement of partners in the planning phases of the research, building on previous mutual experiences and performing preliminary background work on contextual factors and policy processes.

Supporting ongoing collaboration by maintaining dialogues through frequent in-person meetings is essential; however, the work balance between stakeholders may limit these possibilities. Instead, challenges that may arise during the research project can be mitigated by partners setting collaborative goals, strategies, limitations and models of participation in the early stages of the process. The inclusion of a social scientist in the research group may support effective collaboration as they can provide feedback and critical reflections to optimize the research protocol as research evolves.

Developing knowledge translation and exchange is a major issue in collaborative research because it needs to consider disciplinary jargon, which may affect communication of findings between partners and will benefit from a knowledge translation plan being implemented by a team member. A good way to mediate this is to assign one of the collaborators an intermediary role to disseminate results promptly and address any issues related to data ownership before dissemination.

4.6. The Mediterranean Basin: asymmetries between Northern and Southern Mediterranean countries

The Mediterranean basin includes 136 biosphere reserves in 22 countries, which together share a 46,000 km coastline across Africa, Asia and Europe and a rich common history of sea merchants and travelers facilitating trade and cultural exchange between peoples of the region. Today, Mediterranean countries differ concerning their system of governance, level of economic development and commitment to regional integration and fall under multiple sub-regional entities, such as the EU-27, EU candidate countries, the Mashreq (Levant), the Maghreb, the Arab League, the Organization of Islamic Cooperation and the African Union.

4.6.1. *Economic development*

Many Southern Mediterranean economies remain agriculture- and energy-based "rentier states" with a developing service sector, while North Mediterranean economies maintain a strong industrial base and high-tech service industry that promotes a "knowledge society" (Beblawi 1990). While Northern Mediterranean countries are open and export-oriented, many South Mediterranean countries maintain relatively closed and state-controlled economies. North Mediterranean countries are the largest exporters in the world, and some are prime locations for global investment, while Southern Mediterranean countries have consistently captured a low share of world trade, technology exports and foreign direct investment (Abdelkrim and Henry 2009). In fact, although one fifth of the

Organization for Economic Co-operation and Development (OECD) members are Mediterranean states, none of the OECD members are Mediterranean Arab states.

4.6.2. *Human and social development*

Populations in many North Mediterranean countries are aging, while their southern counterparts are experiencing a "youth bulge", with 60% of the Arab population currently below 25 years of age (Johansson de Silva and Silva-Jauregui 2004). North Mediterranean countries are facing skill shortages and mounting fiscal problems, while most Southern Mediterranean countries fare poorly on global human development indexes, and Arab Mediterranean countries face rampant unemployment, food shortages, environmental degradation and water scarcity (United Nations Development Program 2009a, 2009b).

4.6.3. *Trade and economic integration*

Throughout history, the Mediterranean has been central to global trade and commerce; however, more recently the region has turned into an economy characterized by one of the lowest levels of trade integration in the world. Despite attempts to establish regional and sub-regional free trade areas, the level of trade and economic integration amongst the Southern Mediterranean countries remains considerably below the level achieved by other regional organizations, such as NAFTA or ASEAN (World Bank Group 2008). In contrast, the level of trade integration between Southern Mediterranean countries and the EU is more advanced, although it remains shallow and fraught with serious obstacles.

4.6.4. *Scientific contributions and representation*

The discrepancy in scientific publications and quality of research between northern and Southern Mediterranean basin countries is high relative to population size (Sweileh et al. 2015). Scientific works in Southern Mediterranean countries are also less likely to be represented in relevant datasets. For example, one of the most comprehensive compilations of research on biosphere reserves, the *International Journal of UNESCO Biosphere Reserves*, includes more than 2,700 publications, none of which are from the 20 biosphere reserves present in the Levant Arab and North African states. This may give the impression that there is no research on biosphere reserves in these countries and may limit collaboration from interested researchers elsewhere.

4.6.5. *Impediments to collaboration across the Mediterranean*

Remote collaborative activities between countries in the Mediterranean are still difficult despite sophisticated cyber infrastructure, which is not enough to mitigate the complex reality of a region where individuals lack a common vocabulary and working styles and where attempts to store, publish and share data using globally accessible tools are hindered by time restrictions and lack of information (Stuart et al. 2018). Furthermore, conditions needed to allow the necessary space and time for reflection and collaborative efforts are hindered by the rapid pace of today's academic world and the dominant mode of evaluating scientific performance (Guimarães et al. 2017). The limited participation of researchers from Southern Mediterranean countries in global studies is another challenge to a balanced understanding of enablers of and barriers to collaborative research in the Mediterranean. In 2018, Springer Nature published a white paper[2] entitled "Practical challenges for researchers in data sharing" with 7,700 respondents. The findings revealed a discrepancy between Northern and Southern Mediterranean countries because most Mediterranean participants in this global study were from North Mediterranean countries (96.7%). The lack of participation from Southern Mediterranean researchers in this study does not allow one to draw conclusions concerning limitations for data sharing, as well as for collaborative research in general, across the Mediterranean.

4.6.6. *Regionalism and conflict*

While the Mediterranean basin countries share a common history and similar cultural traits, there are few places in the world today that are more politically divided and conflict-ridden. As a result, the region has become home to a unique blend of different Pan-Arab, EuroMediterranean[3], Pan-African and Maghreb projects of region-building that are evolving both in concert and competition. Political and economic orientations and allegiances are pulling Mediterranean countries in different directions, leading to a stifled regional situation that has translated itself into a long list of ethnic groups, territories and countries with limited or disputed recognition, as well as countries with rogue state labels.

---

2 A white paper is an authoritative report or guide that informs readers concisely about a complex issue and presents the issuing body's philosophy on the matter. It is meant to help readers understand an issue, solve a problem or make a decision.

3 Mediterranean regionalism is largely seen as a European Union (EU) project, potentially instrumentally valuable in the southern and eastern Mediterranean but not yet a focus for shared identity (Schoenfeld and Rubin 2004).

4.6.7. *Academic boycotts*

It can be generally stated that the progress of academic work relies on free exchange between researchers, as indicated in Statute 5 of the International Council of Science (ICSU):

> The principle of the Universality of Science (and Learning)[4] is fundamental to scientific progress. This principle embodies freedom of movement, association, expression, and communication for scientists, as well as equitable access to data, information, and research materials. In pursuing its objectives in respect of the rights and responsibilities of scientists, the ICSU actively upholds this principle, and, in so doing, opposes any discrimination based on such factors as ethnic origin, religion, citizenship, language, political stance, gender, sex, or age.

In light of ongoing conflicts that plague the Mediterranean basin, the Principle of Universality clearly stands in tension with the practice of academic boycotts: "the systematic withholding of normal professional relations from academics as a means to achieving some goal, typically either punishment or the bringing about of some change in behavior or policy (Rodin and Yudkin 2010)". Boycott action may include refusing to attend conferences or engage in collaboration, withholding data or results, refusing requests from universities to comment on applications for promotion of university staff, refusing to referee or consider articles submitted to learned journals or refusing applications for university posts.

4.7. Travel limitations

The issuing of travel visas is another threat to collaborative research, as not all researchers in Mediterranean basin countries can travel freely to attend conferences and visit collaborators.

4.7.1. *Language barriers*

The cultures and languages of the Mediterranean are diverse and complex, to the extent that the University of Göttingen offers a degree in "Cultures and Languages of the Mediterranean". In the North Mediterranean basin, there are more languages than there are states and they are divided into regional dialectal groups and further

4 The reasoning that underlies the Principle of Universality in the natural sciences also applies in part (and with appropriate modification) to the social sciences and the humanities. We shall therefore refer to the Principle of the Universality of Science and Learning (or simply the Principle of Universality), which we take to refer to all the academic fields collectively.

subdivided into local dialects[5]. Furthermore, many European countries recognize minority languages and sub-official languages[6]. In Turkey, there are 43 languages, some spoken by as little as 300 people, even though Turkish is the official language of the country (Yagmur 2001).

Modern Standard Arabic (MSA) is the official language of all Mediterranean Arab countries; however, because of Diglossia in the Arabic Language[7], this does not facilitate collaborative researchers among these countries because citizens are more proficient and comfortable in their Arabic dialect subgroups[8] (Al Suwaiyan 2018). As a result, when meetings are organized between researchers of Mediterranean Arab countries, the meeting language is usually a foreign one, namely English or French.

---

5 A dialect can be simply defined as a variety of a language spoken locally and considered to be a less formal and less prestigious variety of that language (Wardhaugh and Fuller 2015), with a distinctive construction and inflectional elements (Milllar and Trask 2015). Over time, these dialects gain different linguistic aspects that diverge from the mother language and from each other. Eventually, these dialects evolve to be different, independent languages (Milllar and Trask 2015). An additional factor that has a role in determining the number of languages in the world is whether to consider the different varieties of a specific language to be different languages or different dialects (Comrie 2009).

6 The European Charter for Regional or Minority Languages (ECRML) is a European treaty (CETS 148) adopted in 1992 under the auspices of the Council of Europe to protect and promote historical regional and minority languages in Europe.

7 Diglossia in general is "a situation in which there are two distinct codes with clear functional separation; that is, one code is employed in one set of circumstances and the other is an entirely different set" (Wardhaugh and Fuller 2015). These codes are related to each other but regarded as different varieties, each used differently depending on the reason and situation. Thus, one may be high and the other low. An example of diglossia is the Arabic language. Diglossia in Arabic is said to be an old linguistic situation that arose during the spread of Islam, when the Arabic language came into contact with other languages and non-Arabs began to speak Arabic. It refers to the existence and use of two or more types of Arabic in an Arabic-speaking country. Of these types of Arabic, one is Modern Standard Arabic, considered to be the high variety, while the other is usually a colloquial variety, which is the low one. Each type is used for a specific purpose (Al-Sobh et al. 2015).

8 Egyptian dialects are mainly spoken in Egypt; however, this dialect is widely known as it is often used in Arab film and television series productions and is also used by many Egyptian teachers in Arab countries (Mitchell 1976). The Levantine dialects are spoken in the northern portion of the Arabian Peninsula, such as in Jordan, Lebanon, Syria and Palestine. These dialects use unique intonations, which make them desirable to many Arabs. The Gulf dialects may be the closest in vocabulary and sound to Modern Standard Arabic. The Maghrebi dialects, spoken in northern Africa, such as Tunisia and Maghreb, were influenced by the French and Berber languages. As such, they are the most difficult dialects to understand among other Arabic speakers.

English is the commonly used language in research settings. Many students in most Mediterranean countries learn English as a foreign language, including countries that used to teach Russian. On the other hand, North African Mediterranean countries teach and use French in research settings. As a result, Mediterranean scholars who are proficient in French and English are at an advantage in forming collaborations without being restricted to either English- or French-speaking partners. Furthermore, this schism in research language use only adds to the costs of collaborative research projects because it requires the services of interpreters in conferences and translations of project documents.

4.7.2. *Institutional structures promoting collaborative research in the Mediterranean*

There are several institutional frameworks and collaborative research centers that seek to promote collaboration, whether directly or indirectly, across the Mediterranean. This section provides a few examples of such initiatives:

– The Association for Medical Education in the Eastern Mediterranean (AMEEMR)'s aim is to develop the highest scientific and ethical standards in medical education and innovative management in the Eastern Mediterranean region through enhanced communication; regular conferences and educational events; exchange of teaching, research and evaluation materials; mobility; and the dissemination of appropriate and evidence-based procedures.

– The International Centre for Advanced Mediterranean Agronomic Studies (CIHEAM) is a Mediterranean intergovernmental organization composed of 13 Mediterranean Member States with headquarters in Paris and four institutes in Bari (Italy), Chania (Greece), Montpellier (France) and Zaragoza (Spain). CIHEAM's objectives are to combat all forms of waste, food and nutrition security, inclusive development and prevention of crises.

– Collaborative Genomics for Human Health and Cooperation in the Mediterranean Region is rooted in the region's demographic history. The center is devoted to genomic analysis of Mediterranean populations to offer unique insights into human evolution and early human migration.

– The Centre for Economic Research on Mediterranean Countries (CREM) at Akdeniz University was established to promote the relationship between Turkey and Mediterranean countries by focusing on free trade agreements, agriculture, energy, tourism and migration issues.

– The Euro-Mediterranean Center on Climate Change (CMCC) is a non-profit research institution established in 2005 with the financial support of the Italian Ministry of Education, Universities, and Research. CMCC manages and promotes scientific and applied activities in the field of international climate change research and represents an institutional point of reference for decision-makers, public institutions and private and public companies seeking technical-scientific support.

– The Euro-Mediterranean Study Commission (EuroMeSCo) is a network of research centers in the Euro-Mediterranean area comprising 104 institutes and including 29 Mediterranean countries with the mission of fostering evidence-based and policy-oriented analysis on Euro-Mediterranean politics and policies through joint research programs, providing a platform for dialogue between researchers and key stakeholders and boosting the role of research institutes in strengthening Euro-Mediterranean relations.

– The Institute of Mediterranean Agricultural and Environmental Sciences (ICAAM) is a research and development unit at the University of Évora that engages multidisciplinary teams to develop knowledge contributing to the sustainability of Mediterranean agriculture and associated ecosystems through efficiency in the use of production factors; the quality and added value of agri-food products and ecosystems integrity and landscape multifunctionality.

– The International Centre on Mediterranean Biosphere Reserves, Two Coastlines United by their Culture and Nature is a center located in Spain that documents scientific research and knowledge on all biosphere reserves in the Mediterranean basin and serves as a platform for training and transferring advanced knowledge on environmental and societal issues between countries. It will facilitate joint programs between biospheres reserves.

– The International Union for Conservation of Nature's Centre for Mediterranean Cooperation's (IUCN-Med) purpose is to influence, encourage and aid Mediterranean societies in achieving sustainable development – while conserving their natural resources and ensuring they are used sustainably.

– The Islamic Cooperation Youth Forum (ICYF) includes many Mediterranean countries and seeks to promote the exchange of ideas and experiences, increase youth involvement in society, promote solidarity among youth organizations of the region, promote international collaborative projects that support solidarity, dialogue and cooperation among youth organizations and address problems faced by youth in these countries.

– The Mediterranean Institute for Environmental Studies' mission is to become a reference center for responsible research and innovation on social, economic and environmental issues in the Mediterranean region. Interdisciplinary research teams focus on sustainable management of natural resources, adaptation to climate changes, sustainable lifestyles, quality and safety food production and nutrition, development of a sustainable and inclusive society and sustainable rural and urban development.

– The Network of Mediterranean Plant Conservation Centers (GENMEDA) is a network of seed banks and conservation centers for genetic resources of the Mediterranean flora. GENMEDA's members from 12 Mediterranean countries contribute to the network through the improvement of human capital, sharing equipment and methodologies to boost applied knowledge, the creation of living plant collections and joint initiatives and encourage of environmental education and public awareness about biodiversity conservation.

– The Organisation Internationale de la Francophonie includes many Mediterranean countries as members or observers and seeks to connect various French-speaking peoples to promote cultural and linguistic diversity, peace, democracy and human rights, education, training, higher education and research.

– The Regional Activity Centre for Specially Protected Areas (RAC/SPA) was established in Tunisia for the protection of the Mediterranean Sea against pollution. The RAC/SPA promotes research and includes coordinating capacity-building and technical support. The RAC/SPA's mission is to provide assistance to Mediterranean countries in the implementation of their commitments under the SPA/BD Protocol.

– The Union for the Mediterranean (UFM) is an intergovernmental organization of 42 member states including 15 Mediterranean partner countries that aims to promote stability and integration throughout the Mediterranean region. The UFM's goal is to increase north–south and south–south integration in the Mediterranean region and to support the countries' socio-economic development through projects that have a direct impact on common regional challenges.

4.8. Conclusion

Collaborative research in the Mediterranean basin is hindered by asymmetries at the social, cultural, economic, political and research capacity levels. On the other hand, Mediterranean countries share a common eco-geographic identity and a natural heritage that has shaped the relationship between people and nature and molded comparable traditions and rural lifestyles around the basin. Biosphere reserves, which capitalize on this relationship between people and their environment, offer an opportunity to "ground" collaborative research that targets problems and capitalizes on opportunities peculiar to such areas. To promote and sustain collaborative research between stakeholders of biosphere reserves, including researchers, practitioners, residents and society at large, there is a need to identify specific motives behind such collaborations that will eventually lead to the formation of a community. For example, similar socio-ecological problems that are addressed through the same approaches and rely on the same skill sets and knowledge may be a motive for engaging in collaborative research with the Mediterranean biosphere reserve community, making it possibe to gain experience from success stories in other Mediterranean biosphere reserves and receive training from researchers and practitioners. Through collaborative research, the potential of Mediterranean biosphere reserves to evolve through similar paths and contribute to the wellbeing of people will, in turn, give biosphere reserve stakeholders the incentive to work together and build friendships driven by a common sense of belonging to a "Mediterranean people", which in turn will sustain collaborations. For such collaborative research to see the light and sustain itself, it will need to address basic elements of collaborative research, such as the provision of adequate funding to support interdisciplinary practices, facilitating data sharing, offering opportunities for and facilitating mobility, ensuring representation of all stakeholders and promoting inclusion by adopting the main Mediterranean languages. Partners in the EduBioMed project have taken small steps towards providing digital support for promoting collaborative research between stakeholders of Mediterranean biosphere reserves through the development of a collaborative digital platform that will serve as a communication platform for the Mediterranean Biosphere stakeholder community; integrate information generated by stakeholders, featuring recent work and publications; and provide links to the various biosphere reserve tools and repositories.

4.9. References

Abdelkrim, S. and Henry, P. (2009). Foreign direct investment in the Med countries in 2008. Report, ANIMA Investment Network.

Al Suwaiyan, L.A. (2018). Diglossia in the Arabic language. *International Journal of Language and Linguistics*, 5(3), 228–238.

Amabile, T.M., Patterson, C., Mueller, J., Wojcik, T., Odomirok, P.W., Marsh, M., Kramer, S.J. (2001). Academic–practitioner collaboration in management research: A case of cross-profession collaboration. *Academy of Management Journal*, 44(2), 418–431.

Anderson, E.E. (2013). Views of academic and community partners regarding participant protections and research integrity: A pilot focus group study. *Journal of Empirical Research on Human Research Ethics*, 8(1), 20–31.

Beaver, D. and Rosen, R. (1978). Studies in scientific collaboration. The professional origins of scientific co-authorship. *Scientometrics*, 1(1), 65–84.

Beaver, D. and Rosen, R. (1979a). Studies in scientific collaboration. Scientific co-authorship, research productivity, and visibility in the French scientific elite, 1799–1830. *Scientometrics*, 1(2), 133–149.

Beaver, D. and Rosen, R. (1979b). Studies in scientific collaboration. Professionalization and the natural history of modern scientific co-authorship. *Scientometrics*, 1(3), 231–245.

Bozeman, B. and Corley, E. (2004). Scientists' collaboration strategies: Implications for scientific and technical human capital. *Research Policy*, 33(4), 599–616.

Bukvova, H. (2011). Scientists online: A framework for the analysis of Internet profiles. *First Monday*, 16(10) [Online]. Available at: https://doi.org/10.5210/fm.v16i10.3584.

Coleman, D.C. (2010). *Big Ecology: The Emergence of Ecosystem Science*. University of California Press, Berkeley.

Escribano, G. (2000). Euro-Mediterranean versus Arab integration: Are they compatible? *Journal of Development and Economic Policies*, 3(1), 23.

Eurostat (2015). Euro-Mediterranean statistics. Report, European Commission, Brussels.

Foe, G. and Larson, E.L. (2016). Reading level and comprehension of research consent forms: An integrative review. *Journal of Empirical Research on Human Research Ethics*, 11(1), 31–46.

Gimpel, J. (2018). Researchers' challenges in sharing data across geographic borders and disciplines [Online]. Available at: https://group.springernature.com/gp/group/media/press-releases/archive-2018/researchers-challenges-in-sharing-data-cross-geographic-borders-/15545272.

Goddard, J.T., Cranston, N., Billot, J. (2006). *Making it Work: Identifying the Challenges of Collaborative International Research*. University of Calgary, Calgary.

Guimarães, H., Fonseca, C., Gonzalez, C., Pinto-Correia, T. (2017). Reflecting on collaborative research into the sustainability of Mediterranean agriculture: A case study using a systematization of experiences approach. *Journal of Research Practice*, 13(1).

Hazem, B. (1990). The rentier state in the Arab World. In *The Arab State*, Giacomo, L. (ed.). Routledge, London.

Hu, C. and Racherla, P. (2008). Visual representation of knowledge networks: A social network analysis of hospitality research domain. *International Journal of Hospitality Management*, 27(2), 302–312.

Johansson de Silva, S. and Silva-Jauregui, C. (2004). Migration and trade in MENA: Problems or solutions for unemployment in MENA? Report, World Bank Group on Unemployment, Washington.

Kervin, K., Cook, R.B., Michener, W.K. (2014). The backstage work of data sharing. *Proceedings of the 18th International Conference on Supporting Group Work*, New York.

Kessler, R. (2018). White paper: Practical challenges for researchers in data sharing. *Learned Publishing*, 31(4), 417–419.

Kibaroglu, M. (2009). *Eastern Mediterranean. Countries and Issues*. Foreign Policy Institute, Ankara.

Koporc, Z. (2018). *Ethics and Integrity in Health and Life Sciences Research*. Emerald Publishing Limited, Bingley.

Lee, S. and Bozeman, B. (2005). The impact of research collaboration on scientific productivity. *Social Studies of Science*, 35(5), 673–702.

Leydesdorff, L. and Wagner, C.S. (2008). International collaboration in science and the formation of a core group. *Journal of Informetrics*, 2(4), 317–325.

Meadows, A.J. (1974). *Communication in Science*. Butterworths, London.

Meleis, A.I. (2011). *Theoretical Nursing: Development and Progress*. Lippincott Williams and Wilkins, Philadelphia.

Melin, G. (2000). Pragmatism and self-organization: Research collaboration on the individual level. *Research Policy*, 29(1), 31–40.

Michener, W.K. (2015). Ecological data sharing. *Ecological Informatics*, 29, 33–44.

Morrison, D. (2003). *E-learning Strategies: How to Get Implementation and Delivery Right the First Time*. John Wiley & Sons, New York.

Nielsen, K.M. (2013). Biosafety data as confidential business information. *PLoS Biology*, 11(3).

Olson, J.S. and Olson, G.M. (2013). Working together apart: Collaboration over the Internet. *Synthesis Lectures on Human-Centered Informatics*, 6(5), 1–151.

Øye, C., Sørensen, N.Ø., Dahl, H., Glasdam, S. (2019). Tight ties in collaborative health research put research ethics on trial? A discussion on autonomy, confidentiality, and integrity in qualitative research. *Qualitative Health Research*, 29(8), 1227–1235.

Özçelik, T., Kanaan, M., Avraham, K.B., Yannoukakos, D., Mégarbané, A., Tadmouri, G.O., Levy-Lahad, E. (2010). Collaborative genomics for human health and cooperation in the Mediterranean region. *Nature Genetics*, 42(8), 641–645.

Pampel, H., Vierkant, P., Scholze, F., Bertelmann, R., Kindling, M., Klump, J., Dierolf, U. (2013). Making research data repositories visible: The re3data.org registry. *PloS One*, 8(11).

Pampel, H., Vierkant, P., Scholze, F., Bertelmann, R., Kindling, M., Klump, J., Dierolf, U. (2014). Making research data repositories visible: The re3data.org registry. *Data Analysis and Knowledge Discovery*, 30(3), 26–34.

Pettit, R. and Prather, J. (2017). Automated assessment tools: Too many cooks, not enough collaboration. *Journal of Computing Sciences in Colleges*, 32(4), 113–121.

Piwowar, H. (2013). Altmetrics: Value all research products. *Nature*, 493(7431), 159.

Qin, J., Lancaster, F.W., Allen, B. (1997). Types and levels of collaboration in interdisciplinary research in the sciences. *Journal of the American Society for Information Science*, 48(10), 893–916.

Ribeiro, B.E., Smith, R.D., Millar, K. (2017). A mobilizing concept? Unpacking academic representations of responsible research and innovation. *Science and Engineering Ethics*, 23(1), 81–103.

Rodin, D. and Yudkin, M. (2010). Academic boycotts. *Journal of Political Philosophy*, 19, 465–485.

Rycroft-Malone, J., Burton, C.R., Bucknall, T., Graham, I.D., Hutchinson, A.M., Stacey, D. (2016). Collaboration and co-production of knowledge in healthcare: Opportunities and challenges. *International Journal of Health Policy and Management*, 5(4), 221.

Schoenfeld, S. and Rubin, J. (2011). Contrasting regional environmentalisms in the eastern Mediterranean: A social constructionist perspective. *Revue en ligne de géographie politique et de géopolitique*, 14.

Shepherd, K.G., Fowler, S., McCormick, J., Wilson, C.L., Morgan, D. (2016). The search for role clarity: Challenges and implications for special education teacher preparation. *Teacher Education and Special Education*, 39(2), 83–97.

Sonnenwald, D.H. (2007). Scientific collaboration. *Annual Review of Information Science and Technology*, 41(1), 643–681.

Stuart, D., Baynes, G., Hrynaszkiewicz, I., Allin, K., Penny, D., Lucraft, M., Astell, M. (2018). Practical challenges for researchers in data sharing. *Springer Nature*, 59750(11), 1.

Sweileh, W.M., Sa'ed, H.Z., Al-Jabi, S.W., Sawalha, A.F. (2015). Contribution of Arab countries to breast cancer research: Comparison with non-Arab Middle Eastern countries. *BMC Women's Health*, 15(1), 25.

Tenopir, C., Allard, S., Douglass, K., Aydinoglu, A.U., Wu, L., Read, E., Frame, M. (2011). Data sharing by scientists: Practices and perceptions. *PloS One*, 6(6), e21101.

Thorsteinsdottir, O. (2000). External research collaboration in two small science systems. *Scientometrics*, 49(1), 145–160.

Trovato, M.G. and Abunnaser, Y. (2015). East Mediterranean landscape observatory. *Landscape Observatories in Europe II*, 56.

Uhlir, P.F. and Schröder, P. (2007). Open data for global science. *Data Science Journal*, 6, OD36-OD53.

United Nations Development Program (2009a). Human development report 2009. Overcoming barriers. Report, UNDP, New York.

United Nations Development Program (2009b). Arab human development report 2009. Challenges to human security in Arab countries. Report, UNDP, New York.

Wine, O., Ambrose, S., Campbell, S., Villeneuve, P.J., Burns, K.K., Vargas, A.O. (2017). Key components of collaborative research in the context of environmental health: A scoping review. *Journal of Research Practice*, 13(2), 2.

World Bank Group (2008). Middle East and North Africa Region: 2008 economic developments and prospects. Report, World Bank Group, Washington.

Yagmur, K. (2001). Turkish and other languages in Turkey. In *The Other Languages of Europe. Demographic, Sociolinguistic, and Educational Perspectives*, Extra, G. and Gorter, D. (eds). Multilingual Matters, Clevedon.

5

Scientific Tourism in Multi-Labeled Protected Areas: The Ecological Transition and Controversy in the Mountains

5.1. Introduction

Historically, the development of tourism in rural areas has been crossed by a tension between land planning and the protection of nature (Marié 1989; Larrère et al. 2009). In the Alps, the (hyper)anthropization of ski resorts coexists with the (multiple) labeling of protected natural areas (Mauz 2005; Larrère et al. 2009). The combination of the exhaustion of the "white gold" model and the precocity of the mountain signs of climate change, particularly evident in the Southern Alps, invites us to reconsider the forms of territorial innovation and their eco-compatibility (Fort 2015; GREC SUD 2018; Soubirou and Jacob 2019). In this context, some authors call for a change of paradigm in favor of an ecological transition for which the mountainous territories could be precursors (Bourdeau 2009; Oiry-Varacca 2012).

At the same time, projects involving the touristic development of scientific culture (re)emerge, which are prompted by local actors concerned with anticipating the "*après-ski*" and, to a greater or lesser degree, advocating for a dynamic ecological transition. They introduce themselves as the expression of an alternative tourist system, "after-tourism", a modality allegedly able to reinvent a concept of tourism more at ease with the societal and environmental challenges of the 21st century (Bourlon et al. 2017). At the same time, these projects reveal distinctive

---

Chapter written by Mikaël CHAMBRU and Cécilia CLAEYS.

forms of tourism practices and products incorporating the scientific dimension in different ways and to varying degrees (Mao and Bourlon 2011). This scientific tourism gives rise to emerging places, which constitute support territories for a scientific culture in the making (Schiele and Jantzen 2003).

This chapter seeks to grasp the socio-environmental issues raised by the touristic development of scientific culture, with a special focus on the case of the Hautes-Alpes[1]. The question underlying this chapter is the following: how does the touristic development of scientific culture in mountain territories reveal the pre-existing tensions between democratic, cultural and environmental issues as a response to the injunction of the ecological transition?

The analysis is structured around three hypotheses relating to the forms of transition brought about by the touristic development of scientific culture from a socio-political, cultural and environmental perspective. The first hypothesis concerns the ability of the promoters of these projects to be part of participatory and inclusive forms of governance. The second hypothesis concerns the accessibility of scientific culture. The third hypothesis tests the compatibility of these tourist development projects with the preservation of biodiversity and landscapes in the context of climate change.

The corpus of qualitative data mainly comprises, on the one hand, of the available technical documentation relating to the project, its (traditional and digital) media coverage and the Internet publications (blogs and social networks) of the protagonists. On the other hand, it includes the semi-structured interviews conducted with the local project leaders and their opponents, as well as field notes compiled between spring 2019 and autumn 2020[2].

The analysis concerns a project for creating a space for scientific mediation, a hotel complex and a cable car on the peak of Château-Renard (2,989 meters), located in the municipality of Saint-Véran. This project was part of a multi-labeled protected natural area, with a location intended to be within the Queyras Regional

1 This work was carried out thanks to the financial support of the Labex ITEM (ANR-10-LABX-50-01) within the framework of the "Investments for the Future" program, managed by the French National Research Agency (ANR). The research project in question is "Mountain, Communication and Astronomy: Public Issues Around Science and Tourism (Maestro)": https://maestro.hypotheses.org.

2 Part of this research work was carried out with the support of Aphélie Savarino during the winter of 2020 as part of an internship at the GRESEC Laboratory and the Labex ITTEM. It gave rise to a second-year master's dissertation defended in June 2020 and entitled "Scientific Culture and Narrative of a Mountain Countryside Territory. The case of Saint-Véran" (Queyras, Hautes-Alpes).

Natural Park (PNRQ), hosting five Natura 2000 sites[3], a national nature reserve[4] and a cross-border biosphere reserve (Mont-Viso) via UNESCO's Man and the Biosphere (MAB) Programme. The Château-Renard site already hosts astronomical science facilities that were established in 1974 and renovated in 2015 (see Figures 5.1 and 5.2).

Regularly advertised as one of the most beautiful in continental Europe, since the end of the 1970s, the sky of the municipality of Saint-Véran has been the object of more or less sustained scientific activities, engaged as a support for the dissemination of knowledge around astronomy and as a lever for touristic development. In this sense, this scientific tourism project is part of the bosom of astrotourism within protected natural areas (Charlier 2018). Supported by the municipality of Saint-Véran, the new tourism development project – with a cost of EUR 25 million – was presented as a possibility for renewing the economic development of the municipality thanks to the revitalization of scientific tourism.

Figure 5.1. *Located at 2,936 meters above sea level, the Château-Renard observatory was renovated in 2015 in order to develop alternative tourist offers promoting the dissemination of astronomical scientific culture (summer view)*

3 The Queyras park is responsible for five Natura 2000 sites: Haut-Guil, Mont Viso, Valpréveyre, Rochebrune, Izoard, Cerveyrette valley, Durancian and Queyrassin steppe, Bois des Ayes and Haut-Guil valley: https://www.pnr-queyras.fr/natura-2000/.

4 Ristolas-Mont Viso.

Figure 5.2. *Equipped with instruments made available by research institutions, the Château-Renard observatory has been welcoming amateur astronomers since 1990, who come from all over Europe to get involved in participatory science (winter view)*

5.2. The ecological transition: from the injunctions to the different socio-political and cultural references

Since the 2000s, the notion of ecological transition has tended to replace that of sustainable development, which had, for its part, been born in the 1980s. The paternity of the notion of transition is attributed to the environmental activist Rob Hopkins, whose research focuses on alternatives to the fossil fuel-dependent model of economic growth (Hopkins 2010).

In France, this concept has undergone a process of institutionalization: in 2013, the French National Council for Ecological Transition was created. Later, in 2017, a ministerial reorganization gave birth to the French Ministry of Ecological and Solidarity Transition, responsible for transport. The notion of transition is also enshrined in French law addressing the energy issue (Law No. 2015-992 from August 17, 2015). It is now included in several national and regional plans[5], among which the most recent to date is the "France Relaunch" plan. Worth a total of EUR 100 billion, this exceptional support program for the economy weakened by the Covid-19 health crisis aims to invest EUR 30 billion in the ecological transition,

5 Like, for example, the climate plan for the Southern region: https://www.maregionsud.fr/planclimat.

aiming to “make ecology the main lever for the recovery and transformation of our economy”[6].

In concrete terms, this amounts to saying that in less than two decades the ecological transition has become a legally framed and financially supported injunction. This process could be derived from an *integration of criticism* as an institutional appropriation/reformulation of concepts born out of protest, which can induce forms of *criticism disarmament* (Chiapello and Boltanski 1999). This process has been particularly emphasized in the analysis of environmental policies (Charvolin 2003). Within the framework of this text, the emphasis is on the integration of the criticism of the transition’s three main challenges: the democratic, the environmental and the cultural questions.

The Rio de Janeiro Declaration of 1992 welded environmental issues to forms of participatory governance. Despite a certain development of the approach over the past few decades, the implementation of this “environmental democracy” comes up against several socio-political difficulties (Claeys and Jacqué 2012). In the field, the scenarios observed are heterogeneous, ranging from the co-construction of the decision associating diverse local populations with forms of “democratic persuasion” to imposition (Claeys and Jacqué 2012; Mermet and Salle 2015). Whether inclusive or exclusive, the question also arises as to the environmental effectiveness of these forms of governance; in other words, as to their effective impact in terms of nature preservation (Salles 2006).

This question becomes crucial with regard to the environmental effectiveness of the ecological transition models. The debate on the compatibility between economic development and environmental protection is inherent in the concept of sustainable development (Lascoumes et al. 2014). While the proponents of degrowth are against the notion that economic development cannot be inherently sustainable, its defenders advocate for a possible compromise. In its initial militant conception, the notion of ecological transition proposes to opt for a change of societal model, breaking with the market economy and aiming at the reduction of social and environmental inequalities (Hopkins 2010).

However, the institutionalization of the concept tends to associate it with the “green growth” principle, as evidenced, for example, by Law No. 2015-992 from August 17, 2015, which regulates the energy transition towards green growth. According to the OECD[7], “green growth means fostering economic growth and development while ensuring that natural assets continue to provide the resources and environmental services on which our well-being relies. To do this it must catalyze

6 See: https://www.ecologie.gouv.fr/france-relance-transition-ecologique.

7 See: http://www.oecd.org/fr.

investment and innovation which will underpin sustained growth and give rise to new economic opportunities".

These injunctions to the ecological transition are part of different socio-political and cultural references. As a corollary, there is the question of the knowledge summoned and its legitimacy. A doctor in social sciences who reconverted to permaculture and was co-founder of the first ecovillage officially in transition, Rob Hopkins associates the ecological transition with forms of promotion of citizen initiatives and local knowledge. The institutionalized ecological transition, on the other hand, tends to promote more technical knowledge.

The question of knowledge on the ecological transition and its legitimacy is particularly exacerbated within the framework of the analysis of scientific tourism projects. In this case, knowledge is both the object and the issue. As knowledge is mobilized to justify the scientific tourism project, it is simultaneously the object of the touristic development and also involves the issue of (non-)appropriation by populations and visitors. In the mountains, scientific mediations have been present for a long time in order to promote natural sites and tourist sites. Articulated around the sciences (astronomy, biology and geology in particular), this tourism may be patrimonial, industrial, technical or environmental and is materialized via scientific mediation and popularization mechanisms. These constitute spaces for singular "science/society encounters" and are driven by communication paradigms that shape the roles and relationships with the public (Chavot and Masseran 2010). These singular spaces have been assigned different missions: sharing knowledge, promoting science, valuing territorial specificities, recruiting future scientists, diversifying the tourist offer, engaging the public in research, justifying socio-political decisions, protecting mountain areas, inducing behavioral changes, etc. In this sense, the touristic development of scientific culture in the mountains contributes to reinforcing the ambiguity of the notion of scientific mediation and its social issues (Bergeron 2016).

5.3. The trajectories of governance forms for a scientific tourism project

The first hypothesis tested by our field observations concerned governance forms as possible transition factors. The aim was to test the capacity of the Saint-Vérannais scientific tourism development project to promote participatory and inclusive forms of governance. The analysis reveals that the project is part of a reactive, if not corrective, concertative trajectory (Claeys and Jacqué 2012). This project, at the time supported by part of the municipal majority, was initially part of a top-down decision-making process centered on the representative bodies and their notables network, without engaging public participation. The split in the municipal majority in relation to this issue acted as the starting point for public awareness and the controversy that ensued. In September 2018, the first public meeting organized in Saint-Véran on the subject was on the initiative of the second deputy mayor, who was opposed to the project.

The protest was then structured around a core of local associative actors who had former experience in mobilization techniques. The latter quickly activated their regional and national networks and the Saint-Véran collective was thus created. Combining different traditional and digital mobilization tools[8] (Granjon 2018), the collective succeeded in persuading the project leaders of applying more inclusive governance forms. In May 2019, the second public meeting organized – this time on the initiative of the mayor – was timidly part of such logic. That meeting was conceived as an information meeting by its organizers. The intention was to convince the public of the merits of the project rather than to consult them or to concert a decision. Even though the inhabitants were not explicitly given the floor, some of them took the initiative and expressed themselves loud and clear during this lively meeting. The accusations of the forcible imposition of an unconcerted project were recurrent and, as a corollary, there were demands for the involvement of the population and the territory actors in the decision-making process.

Widely disseminated on militant social networks from April 2019, the petition against the project obtained 13,822 signatures. In August 2019, a small demonstration accompanied the delivery of a letter from the collective to the mayor asking for a break in the decision-making process and a public debate on the project. The controversy surrounding this tourist development project was at the heart of the spring 2020 municipal elections. Candidacies emerged either to support or to contest the scientific tourism project. In a late reversal, surprising the candidates themselves, the second round of the municipal ballot resulted in the majority of them opposing the project. The former second deputy, dissident, thus became the new mayor of the town.

The governance of this scientific tourism development project experienced a trajectory oscillating between representative and participatory democracy, and as a corollary, between closed sessions and publicity, between ballots and digital social networks. As it happens with many other challenging issues, these two spheres are opposed and intertwined (Blatrix 2009). As the internal debate within the municipal team was prevented by the activation of a decision-making process led by the notables, such a process had to be counterbalanced by an associative mobilization activating highly visible militant networks in the digital public space. There were many comings and goings between the (non-)internal debate at the municipal level and the external debate sparked by an elected municipal official.

8 Public meetings, demonstrations, blog, online petition, etc.

Thus, the dispute over the land planning project arose within the representative bodies, but could only be fully expressed by resorting to digital participatory processes. The victory of those opposing the project was consecrated during the "penalizing" municipal elections. Throughout this controversy, the second deputy played the role of "smuggler" between the representative and participatory bodies. It was in the digital public space that one of the first decisions of the new municipal council was broadcast to abandon the land planning project and to set in motion reflection on the future of the municipality, on its "transition".

The new mayor of Saint-Véran announced:

> We have to move on. Abandoning the cable car project is one thing. But talking about our future is another [...] We are going to rework the Local Urban Plan (LUP) which was under revision, and we are going to try to equip ourselves with the best arguments for the future of Saint-Véran [...]. We won't have big projects such as the cable car. Anyway, these big projects are outdated, unreasonable, both economically and environmentally, we have to 'switch software' regarding our economic development (Radio/TV DICI, September 17, 2020).

This participatory trajectory was both classic and unique. Contesting notable-conducted governance through a network of associative mobilizations which then invest municipal power has become a classic of what used to be called the "*new* social movements", which are now stable in the long term (Chabanet 2020). On the other hand, the singularity of that scenario was due to the fact that (I) the dispute arose from within the municipal authorities; (II) it overflowed those authorities, mainly in a dematerialized form, with a few strong face-to-face moments which ended up in discursive verbal confrontations; and (III) it tipped the municipal elections.

5.4. The ambiguities related to the touristic development of scientific culture

The second hypothesis concerns what is at stake in the accessibility of scientific culture. To what extent do the scientific mediations pervading the project fall within the scope of cultural transition forms? As soon as it was made public, the project was first justified by its promoters as being necessary for the development of scientific activities in the territory. The challenge was then to accommodate the new robotic telescope from the Paris Observatory (CNRS) in Saint-Véran, a facility whose goal was to promote astronomical and astrophysical research, with a strong international dimension in collaboration with American and Chilean astronomers. This telescope was presented as "the largest robotic telescope in Europe at such an altitude" and the result of a political will at the service of a scientific project (NGE et al. 2019). This project would constitute a significant change of scale: since the end of the 1970s, the sky of the municipality had been the theater of scientific activities carried out by *amateur* astronomers with a participatory science logic.

Before trying to launch this new land planning project, a feasibility study financed by the Provence–Alpes–Côte d'Azur[9] region was initiated in 2017. The study concluded on the need to create a cable car to facilitate access for researchers to the observatory from the village center. During the public meeting organized by the local representatives who promoted the project, a researcher from the Systèmes de référence temps-espace (SYRTE) laboratory from the Paris Observatory was present at their side. After the mayor's introductory speech, the SYRTE researcher was the first speaker of the evening, in the role of expert. His proposal was structured around two strong arguments. The installation of a new high-performance telescope at the Château-Renard peak would give French astronomy the means (I) to strengthen its international influence and (II) to invest in the essential popularization of science. He introduced the installation of the cable car as a sine qua non condition for the deployment of the new telescope. Scientific activity and the implementation of a new mode of transport were thus presented as inseparable. The legitimacy of his expert status was intended to contribute to the project's credibility. On the one hand, the idea was to forge alliances with specific audiences (local authorities, investors, etc.) in order to develop the best possible conditions for the project's realization and, on the other hand, to create a movement in favor of the popular support for the project. Science was used for building a trustworthy and reliable actor's legitimacy in order to confront critical audiences and as a means for justifying decisions which escaped the sole logic of the scientific field, reflecting two proven communication paradigms in the relations between science and society (Chavot and Masseran 2010).

A second goal was displayed by the promoters: by publicizing its presence on the territory, the scientific project was portrayed as an opportunity for developing tourism in the name of science. The touristic development of scientific culture was defended as logical and self-evident – and therefore immune to controversy – because Saint-Véran had had a complex for the dissemination of scientific culture supported by the Paris Observatory since 2015: an astrotourism offer at the observatory and a museum space devoted to the sun in the village. The local actors interviewed testified to the great success of the first goal with the public, whereas the second goal did not manage to find its audience among the tourists. The most critical actors called into question the potential tourist attraction of this facility located at the foot of the alpine ski slopes. Science and scientific culture were presented as a potentially marketable "product" on the material and symbolic level, reflecting the most

---

9 After the last territorial reform (2016), France was divided into 13 metropolitan regions, including the current Provence–Alpes–Côte d'Azur region: https://www.maregionsud.fr/.

traditional paradigm underlying science/society relations (Chavot and Masseran 2010).

The specificity of this project compared to other forms of scientific mediation was that its commercial valuation took place in the form of a tourist product in the mountains enabling people to live a personal experience in cultural decentration (Kramar 2011). The goal was to make science attractive and entertaining for tourists, thus contributing to the diversification of the territory‘s tourist offer. In order to make it look more appealing, science was presented under different facets in Saint-Véran, sometimes spectacular (spending a night in an observatory at an altitude of 3,000 meters to observe the sky), sometimes as a fun experience (performing chemistry-related scientific experiments). In both cases, the benefits of science were celebrated, whether in its usefulness, its ethics or the responsibility values it portrays. The promoters of the project deemed it necessary to educate the public in tools for and keys to science and scientific research currently in use, accompanying the popularization of scientific research at the observatory and/or the different foundations which subsidize the sun house.

This way of understanding the dissemination of the scientific culture corresponds to the deficit model, envisaged as the linear transmission of information from experts to the public. This model postulates that the proper transmission of information enables a reduction in the knowledge deficit and that such a reduction contributes to better societal decisions and a strong support for science (Brossard and Lewenstein 2010). In Saint-Véran, this would give credibility to the scientific approach and its actors, while in return, the latter would make it possible to legitimize the new project for the installation of the observatory in the name of the necessary weaving between science/society relations.

The bottom line of this project was more the promotion of a brand image of the municipality associated with science rather than a disinterested sharing of knowledge, reducing “commendable initiatives to an essentially apologetic and propagandistic aspect” (Levy-Leblond 2008). This can be explained by the sociological characteristics of the promoters of the project and their vision on the benefits of science and the virtuous circle set in motion by the impetus for the dissemination of scientific culture. “The increased public recognition of science benefits science itself, scientific bodies, scientists, citizens and even entire nations” (Christensen 2006).

The touristic development of scientific culture at Saint-Véran tends to replay the established modalities for publicizing science and the accompanying popularization logics. In point of fact, these widen the gap they presuppose between science and society by sanctifying science and the knowledge produced, considering that this should be made visible to a large fraction of the public whose lack of knowledge is prejudged and deplored (Bensaude-Vincent 2010). By presenting itself as the place required by the public thirsty for science, the scientific tourism project precisely aims to come to fill this "gap". In the context of the ecological transition, it aims to offer this public new forms of tourism which are not closely related to recreational and outdoor activities. It therefore seems a reformulation of the project for publicizing science already deployed for several decades rather than a social evolution indicating that this publicity is actually changing (Paillart 2005).

5.5. The environmental paradoxes of a scientific tourism project

The third hypothesis explores the capacity of scientific tourism to be part of the ecological transition or even to be its driving force in the mountain territories. The goal in this section is to understand to what extent the Saint-Véran project was (or not) part of the search for a tourist alternative matching the ecological transition principles. In that case, what would be its conception of the ecological transition? Whether fully in line with an ecological transition register or not, did the project meet the requirements for a low environmental impact?

The promoters of the project considered scientific tourism as an opportunity for territorial development to respond to the increasing irregularities in the snow cover of alpine ski areas. As a matter of fact, the small family ski resorts of Queyras particularly suffered from the increasing irregularities in the snow cover. The lowest resort in altitude, located in the town of Aiguille, had already abandoned its alpine skiing activity altogether. In that sense, the Saint-Véran scientific tourism development project was clearly part of a strategy for adapting to climate change. However, does it make part of an ecological transition process?

In the project's feasibility study, the term "transition" appeared only once, referring to the ministry of the same name as a signatory for the extension of the classification of Queyras as a regional natural park. This information was mentioned in the "constraints" part of the study. The consideration of the environmental impact by the feasibility study had not been mentioned as a goal in itself but rather as a regulatory constraint. Still, despite being located in a multi-labeled area, the project was relatively unconstrained by various protection tools. The village of Saint-Véran – departure station for the cable car project – was located in the buffer zone[10] of the

10 In general, the buffer zone within biosphere reserves aims to reconcile economic development with environmental preservation, whereas the central area is dedicated to the conservation of biodiversity (Beuret 2011). However, this biodiversity reservoir designation is above all an international label rather than a reserve with a specific status. While biosphere reserves rely on existing national tools, France favors a flexible approach, allowing them to be adapted to the local conditions and needs (Comba 2011).

Mont-Viso cross-border biosphere reserve. The Château-Renard peak (arrival station) is set only a few kilometers from the central zone.

In this context, the project's environmental compatibility was limited to a minimal integration within the regulatory framework. To do this, the project invoked the eco-efficiency register, searching for technical solutions aimed at minimizing the ecological impact of an activity that generates economic growth (Martinez-Alier 2014). This eco-efficiency was condensed in the futuristic architecture chosen for the high-altitude buildings, inspired by the cinematographic imagination of space stations. The eco-compatibility of the cable car serving the high-altitude tourist area was also highlighted, emphasizing the project's limited footprint in the natural environment, limiting the destruction or degradation of natural and species habitats (NGE et al. 2019).

During the public meeting organized by the promoters of the project, the design office presented the cable car as having low visibility in the landscape thanks to the shape and color chosen for its pylons. However, the argument did not convince the opponents. Not only did they mention the project's lack of eco-compatibility from a material point of view but also in terms of environmental ethics (Larrère 2006). The direct impacts denounced by the opponents to the project concerned the landscape, biodiversity and agropastoralism. The impact provoked by public attendance was, for its part, a hot topic for discussion. The feasibility study was particularly explicit on that point: the project aimed to increase the tourist appeal of Saint-Véran by attracting a growing number of visitors.

To make the initial investment profitable, the project targeted 95,000 visitors per year, for whom additional parking areas were planned. Thus, the project heavily relied on tourism based on individual motorized transport. However, the impact assessment stressed that motorized transport between the home and the visited site was the cornerstone of the touristic transition (Gössling et al. 2002). Those opposing the project were also worried about the deterioration risks in the villagers' quality of life in the face of an increased influx of visitors. These fears were corroborated by the analyses carried out for several decades on the hosting capacities of natural tourist areas. While the authors struggled to agree on the most efficient methods and indicators, they were nonetheless unanimous on the need to take into account both the ecological and psycho-sociological dimensions of the increases in visitor flows (Ginelli et al. 2021).

The cable car retained by the feasibility study had an estimated operating speed of up to 12.5 meters per second, with cabins fitting 30 to 200 people. That choice of speed and high fitting capacity were justified by a desire to democratize the access to the high mountains.

> The cable car linking the village and the plateau of Château-Renard allows everyone to access the high mountains in complete safety, from the contemplative who will use a discovery lift, to the sportsman who will access the mountains either in winter or in summer to discover new sensations, as for the families for whom the departure from Château-Renard will signify a real gateway to the high mountains and its routes within the valley and a descent back to the village (NGE et al. 2019).

However, it is clear that tourism and recreational practices remained socially selective, despite their massification (Daudey et al. 2016; Deldrève and Claeys 2016). The opponents to the project were concerned about two related effects resulting in both a concentration of visitors at the high-altitude tourist site and their dispersion in the surrounding area. In addition to the access to scientific tourism, the cable car would de facto significantly enlarge the ski area, in flagrant contradiction to the charter of the Regional Natural Park of Queyras.

By a "taxonomic acrobatics"[11], the celestial vault was evoked by the promoters of the project as the ultimate naturalness, superior to terrestrial naturalness. This was the argument of the feasibility study's lyrical preamble:

> [...] In the middle of the day, the vast mineral majesty of the surrounding peaks and glaciers inspires our fragile beings to measure the tiny proportions of humanity in the face of telluric forces. But it is when the evening comes that the real vertigo begins on Château-Renard's peak. Looking towards deep black sky, the show of the immeasurable mystery of our universe is offered to our eyes. Myriads of infinitesimally vibrating light signals are calling to us. The celestial vault unfolds and takes us away. Holding its vision with the naked eye is staggering, disarming (NGE et al. 2019).

It is as if, in a reversal of perspective, the cable car and its observatory were the access route to the most majestic wilderness. In response, the opponents denounced the gigantism of a project in contradiction with environmental ethics and the quality of life they aspired to.

5.6. Conclusion

This case study reveals that it is difficult to break free from an anthropocentric tourism model based on technical development and the promotional vision of

11 The expression was coined by Boltanski (1979).

science and scientific culture. It shows the ambivalence of scientific tourism in a transition context at the rural mountain territories, caught between two social logics relying on differing values, issues and purposes: (I) tourism diversification through scientific culture and (II) broadening of scientific culture by transforming it into a touristic offer.

The analysis of the touristic development of scientific culture at Saint-Véran showed little or belatedly participative decision-making. The whole process was exposed to local disputes calling into question the pyramidal logic in public decision-making and the dissemination of knowledge, as well as the low eco-compatibility of land planning developments in the face of growing economic and climatic vulnerabilities. While the protagonists of the controversy agreed on the need to reduce the municipality's economic dependence on "white gold", two development models were opposed. The promoters of the scientific tourism project remained anchored in anthropocentric ethics, summoning classic models of economic growth rather than those relating to the ecological transition (and this not even in the ministerial sense of green growth). While they were keen on emphasizing the project's eco-compatibility, this was a legally constrained and politically reactive posture. On the other hand, those contesting the project mobilized environmental ethics, evoking the protection of natural environments, the defense of agropastoralism and the demands for a better quality of life. These protests denounced a new stage in the territory‘s urbanization, inappropriate for the preservation of its natural patrimony and not providing a satisfactory response to the adaptation to climate change.

This case study exposes the tensions between democratic, cultural and environmental issues inherent in the touristic development of scientific culture in mountain territories. It reveals the prevalence of issues related to territorial attractiveness and the economic benefits targeted by the promoters of these projects. The case of Saint-Véran is all the more eloquent as the scientific tourism project was set in a multi-labeled territory. The scientific council of the regional natural park certainly issued a negative opinion, but one which had no coercive power over the project. Nor did the biosphere reserve buffer zone where the project was set suffice to restrain its proponents. As a last resort, the defense of the territorial transition‘s environmental vision was promoted by the local militant opposition and concretized by its access to municipal power, giving it the keys to the regulatory tool which has until now proven to be the most effective against this development project: the revision of the Local Urban Plan.

5.7. References

Bensaude-Vincent, B. (2010). Splendeur et décadence de la vulgarisation scientifique. *Questions de communication*, 17, 19–32.

Bergeron, A. (2016). Médiation scientifique. *Arts et Savoirs*, 7.

Beuret, J.E. (2011). Quelles voies pour la participation du public à la conservation de la nature ? Le cas des réserves de biosphère. *Développement durable et territoires*, 2, 3.

Blatrix, C. (2009). La démocratie participative en représentation. *Sociétés contemporaines*, 74(2), 97–119.

Boltanski, L. (1979). Taxinomies sociales et luttes de classes. La mobilisation de la "classe moyenne", l'invention des "cadres". *Actes de le recherche en sciences sociales*, 29, 75–105.

Bourdeau, P. (2009). De l'après-ski à l'après-tourisme, une figure de transition pour les Alpes ? Réflexions à partir du cas français. *Revue de géographie alpine*, 97(3).

Bourlon, F., Bourdeau, P., Michel, F., Inostroza, G. (2017). Le tourisme scientifique, un après-tourisme en Patagonie ? *Études caribéennes*, 38, 37–38.

Brossard, D. and Lewenstein, B. (2010). A critical appraisal of models of public understanding of science: Using practice to inform theory. In *Communicating Science: New Agendas in Communication*, Kahlor, L. and Stout, P.A. (eds). Routledge, New York.

Chabanet, D. (2020). Nouveaux mouvements sociaux. In *Dictionnaire des mouvements sociaux*, Fillieule, O. (ed.). Presses de Sciences Po, Paris.

Charlier, B. (2018). "Vous connaissez les Pyrénées le jour, découvrez-les la nuit...". Réflexions autour de l'artialisation in visu des paysages célestes nocturnes pyrénéens. *Revue de géographie alpine*, 106(1).

Charvolin, F. (2003). *L'invention de l'environnement en France*. La Découverte, Paris.

Chavot, P. and Masseran, A. (2010). Engagement et citoyenneté scientifique : quels enjeux avec quels dispositifs ? *Questions de communication*, 17, 81–106.

Chiapello, E. and Boltanski, L. (1999). *Le nouvel esprit du capitalisme*. Gallimard, Paris.

Christensen, L.L. (2006). *The Hands-on Guide to Science Communicators: A Step-by-step Approach to Public Outreach*. Springer, Berlin.

Claeys, C. and Jacqué, M. (eds) (2012). *Environmental Democracy Facing Uncertainty*. Peter Lang, Brussels.

Comba, D. (2011). Le régime légal des réserves MAB en France. *Revue juridique de l'environnement*, 36(3), 297–304.

Daudey, E., Hoibian, S., Lautié, S. (2016). Les Français dépensent près de 5 % de leur budget pour les vacances [Online]. Available at: https://www.credoc.fr/publications/les-francais-depensent-pres-de-5-de-leur-budget-pour-les-vacances.

Deldrève, V. and Claeys, C. (2016). Social inequalities within protected natural areas: An international perspective. *Desenvolvimento e meio ambiente*, 38, 5–8.

Fort, M. (2015). Impact du changement climatique sur les dynamiques des milieux montagnards. *Revue de géographie alpine*, 103(2).

Ginelli, L., Deldrève, V., Claeys, C., Thiann-Bo Morel, M. (2021). *Réguler les usages au nom de leurs impacts. Principes et sentiments d'injustice dans deux parcs nationaux français.* Peter Lang, Brussels.

Gössling, S., Borgström Hansson, C., Hörstmeier, O., Saggel, S. (2002). Ecological footprint analysis as a tool to assess tourism sustainability. *Ecological Economics*, 43, 2(3), 199–211.

GRED SUD (2018). Impacts du changement climatique et transition(s) dans les Alpes du Sud. Notes of the working group "Montagne", GREC SUD.

Granjon, F. (2018). Mouvements sociaux, espaces publics et usages d'Internet. *Pouvoirs, revue française d'études constitutionnelles et politiques*, 164, 31–47.

Hopkins, R. (2010). *Manuel de transition : de la dépendance au pétrole à la résilience locale.* Éditions Écosociété, Montreal.

Kramar, N. (2011). Le tourisme scientifique en question : vers de nouvelles potentialités. *Tourisme et apprentissages*, 95–100.

Larrère, C. (2006). Éthiques de l'environnement. *Multitudes*, 24, 75–84.

Larrère, R., Bernadette Lizet, B., Berlan-Darqué, M. (2009). *Histoire des parcs nationaux : comment prendre soin de la nature ?* Quae, Versailles.

Lascoumes, P., Bonnaud, L., Lebourhis, J.P. (2014). *Le développement durable : une nouvelle affaire d'État. L'écologie en questions*. PUF, Paris.

Levy-Leblond, J.M. (2008). (Re)mettre la science en culture : de la crise épistémologique à l'exigence éthique. *Courrier de l'environnement de l'INRA*, 56, 7–16.

Mao, P. and Bourlon, F. (2011). Le tourisme scientifique : un essai de définition. *Téoros*, 20(2), 94–104.

Marié, M. (1989). *Les terres et les mots*. Méridien-Klinsieck, Paris.

Martinez Alier, J. (2014). *L'écologisme des pauvres. Une étude des conflits environnementaux dans le monde*. Institut Veblen, Paris.

Mauz, I. (2005). Histoire et mémoires du parc national de la Vanoise : trois générations racontent. *Revue de géographie alpine*, 1, 178.

Mermet, L. and Salle, D. (2015). *Environnement : la concertation apprivoisée, contestée, dépassée ?* De Boeck Supérieur, Brussels.

NGE, STEM, MAP, MAW (2019). OSV 2930. Étude de faisabilité. Report. NGE, STEM, MAP, MAW.

Oiry-Varacca, M. (2012). Préface du numéro thématique “Fabriquer des ressources territoriales pour renouveler l’offre touristique dans les Alpes et les Pyrénées”. *Revue de géographie alpine*, 100(2).

Paillart, I. (2005). *La publicisation de la science. Exposer, communiquer, débattre, publier, vulgariser*. Presses universitaires de Grenoble, Grenoble.

Salles, D. (2006). *Les défis de l’environnement : démocratie et efficacité*. Syllepse, Paris.

Schiele, B. and Jantzen, R. (2003). *Les territoires de culture scientifique*. Presses de l’Université de Montréal, Montreal.

Soubirou, M. and Jacob, L. (2019). Quand les montagnes nous invitent à repenser l’innovation sociale. *Revue de géographie alpine*, 107(2).

Spandre, P., François, H., Verfaillie, D., Pons, M., Vernay, M., Lafaysse, E., George, M., Morin, S. (2019). Winter tourism under climate change in the Pyrenees and the French Alps: Relevance of snowmaking as a technical adaptation. *The Cryosphere*, 13, 1325–1347.

Part 2

Educational Practices Relating to Biosphere Reserves: Balance and Prospects

Introduction to Part 2

The second part of this book looks at some of the main focuses of biosphere reserves, namely environmental education, sustainable development, biodiversity, patrimony and territories. The chapters explain topical issues concerning educational prescriptions and practices relating to biosphere reserves in the Mediterranean sphere. Calling into question the links between biosphere reserves and formal education systems, Melki Slimani, University of Kairouan; Angela Barthes, University of Aix-Marseille; and Jean-Marc Lange, University of Montpellier, reflect on the transfer of political skills via environmental and development education in academic curricula. For her part, Sylviane Blanc-Maximin, University of Aix-Marseille, discusses the educational opportunities presented by the partnerships between protected areas and teachers in the formal sector in terms of collective skills enhancement.

In biosphere reserves, agricultural education undoubtedly plays a central role for rural areas. It is in this sense that Nina Asloum, Guillaume Gillet and Laurent Bedoussac, all three from the ENSFEA in Toulouse, explain to us the challenges of agricultural education in the sustainable management of French biosphere reserves. Véronique Chalando and Angela Barthes from University of Aix-Marseille take the question a step further and discuss the opportunities for agricultural education in biosphere reserves, instructing people how to produce differently. Salma Itsmaïl and Bruno Garnier from the University of Corsica return to the tension between tourism and the protection of biodiversity and explain to us the issue of the instrumentalization of education in sustainable development at the service of tourism, as in the case of the Moroccan argan tree biosphere reserve. Saïd Boujrouf and Abdullah Aït L'Houssain, from Cadi-Ayyad University in Marrakech, also discuss the argan tree biosphere reserves, presenting their vision on education and mediation as the tools for forging an alliance and strengthening communication strategies between education and tourism actors.

Introduction written by Angela BARTHES.

6

Teaching How to Produce Differently at a Biosphere Reserve

6.1. Introduction

In a few decades, "modern" agriculture has profoundly transformed agricultural practices in Europe in order to meet production targets. The intensification of cropping systems, characterized by the massive use of synthetic chemical inputs, has generated numerous biocide pollution phenomena and led to the contamination of water nutrients. The drainage of wetlands and/or land re-allotment, as well as the reduction of the agricultural landscape mosaic, are generating a rapid and spectacular decline in biodiversity (Bianchi et al. 2006) and the accompanying ecosystem services (Altieri and Nicholls 2004).

According to the report from the French Ministry of Agriculture (2014), agriculture appears as the second largest GHG[1] emitter after transport, with a total of 20.9% of emissions in 2011. The Food and Agriculture Organization (FAO) (2014) estimates regarding GHGs show that emissions from agriculture, forestry and fisheries have almost doubled over the past 50 years and could continue to increase by 30% by 2050.

These findings undermine an agriculture already badly shaken by repeated food and health scandals since the 1990s (such as "mad cow disease"). In addition, there is the resistance of bio-aggressors and the bundles of presumptions concerning the negative effects of biocides on human health.

In the near future, the ever-increasing demand for foodstuffs, positively correlated with the rise in the world's population, will increase the need for a gain in

---

Chapter written by Véronique CHALANDO and Angela BARTHES.

1 Greenhouse gas.

production and for rethinking the sharing of foodstuffs and consumption. The stakes are high: "boosting" agricultural production, reducing food waste, minimizing the impact on the environment (Tscharntke et al. 2012) and human health and anticipating changes in regulatory constraints related to the use of synthetic plant protection products.

This redefinition of societal goals beyond the sole dimension of production calls for a "paradigm shift in knowledge production and management" (Girard 2014). This is what the law for the future of agriculture, food and forests (2014) aims for, in particular via the "agroecological project for France". It is accompanied by the "teaching how to produce differently" program, aimed at promoting agroecology in agricultural education. It groups a series of national action programs from the Ministry of Agriculture (National Ambition Bio 2017 Program, Sustainable Beekeeping Plan, etc.). Teaching how to "produce differently" resonates with the Sustainable Development Goals (Unesco 2014) and education for sustainable development (ESD), which takes into account this multifactorial environment.

Our chapter presents the curricular challenges of teaching how to "produce differently" and aims at building an analysis model for a coherent agroecological literacy. This analysis is based on a case study: how partnerships between biosphere reserves and agricultural colleges promote coherent teaching for a territorial agroecological transition (Landel and Girard 2020).

6.2. Curricular challenges of teaching how to "produce differently"

To demonstrate the curricular challenges of teaching how "to produce differently", we first focus on the sociology of the school curriculum and the methodologies for elaborating curricular models (Barthes et al. 2021). Then, we will define the challenges for an agroecological literacy. This approach can help us define the criteria that should be retained for a coherent agroecological literacy, which are illustrated by the study of a practical case.

6.2.1. *Curriculum development model*

Here, the term "curriculum" is understood as the planned and coherent intention of an education or training course (Audigier et al. 2006). A curriculum can be defined locally, vertically or sometimes in a mixed way. This point should particularly hold our attention in the context of our study.

There are several ways to control the curriculum, and they are eminently political. In order to understand what is at stake in the structuring of a curriculum, it is worth recalling the different constituent elements of curriculum development models. To grasp the challenges of how to "produce differently", sociologists of the curriculum such as Forquin specify that it is appropriate to define reference materials (Forquin 2008) in order to achieve consistency, whereas Martinand and Lebeaume state that curricula should include contributing knowledge and constitutive practices (Lebeaume 2019). Barthes, Lange and Roth indicate that, in the current globalized context, it is appropriate to also take into account the question of values, norms and injunctive political prescriptions, which are sometimes implicit (Roth et al. 2020). Finally, Alpe specifies that it is necessary to understand the modes of legitimation that contribute to the selection of knowledge and practices, which depends on the influence of actors and institutions.

Briefly stated, the curriculum is an object located at the crossroads between practices and values, sometimes with a political nature and contributing knowledge. It is the influence of these different curricular components that is analyzed here. Our case study focused on agroecological teaching in an agricultural high school closely related to a biosphere reserve. We use the analysis grid developed by Barthes et al. (2021) in order to apply it to a coherent agroecological literacy.

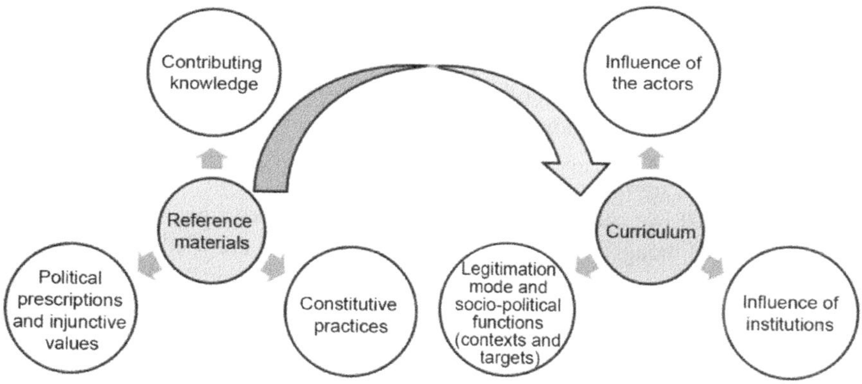

Figure 6.1. *Curriculum development model in a globalized context (Barthes et al. 2021)*

This approach to the socio-politics of the curriculum raises a number of questions:

– What are the reference materials and the reference fields used? What constitutes the contributing knowledge, the constitutive practices, the implicit values and norms and the injunctive political prescriptions?

– Which socio-political functions and which audiences do these elements target? Which modes of legitimation are these choices based upon?

– What is the influence of the actors and institutions, and which ones are involved?

Posing the socio-politics of the curriculum contributes to figuring out what is happening in curriculum development strategies – including the often forgotten dimension involving "political prescriptions and injunctive values" – and understanding the various challenges.

6.2.2. *Distinction between agroecology/agro-ecology: the question of the reference materials*

In the training reference material for the STAV[2] technological baccalaureate, it is stated that: "This training aims to make students capable of summoning knowledge and lived experiences and putting them into perspective by jointly taking into account environmental, social and economic performance in the five technological fields (land planning, production, services, processing, agricultural equipment) available in this baccalaureate branch. This approach, in conjunction with agro-ecology, highlights the complexity of organization, production and decision-making systems, beyond the simple knowledge of models." The prescribed curriculum remains largely vague, and the injunction recommends only a global approach.

In the booklet *L'agro-écologie dans l'enseignement agricole : une dynamique et des résultats* (2019), we can read that: "Agro-ecology implies a major modification in the frameworks of thought, in the modes of acquisition of knowledge and practices. Agricultural education plays a central role in this transformation." Note that the term is written with a hyphen. According to Law No. 2014-1170 from October 13, 2014:

> Public policies aim to promote and sustain agroecological production systems [...] These systems favor the autonomy of agricultural exploitation and the improvement of its competitiveness by maintaining or increasing economic profitability, by improving production's added value and by reducing the consumption of energy, water, fertilizers, plant protection products and veterinary medicinal products, in particular, antibiotics, etc.

According to Alteri, agroecology is "a science which designs productive and sustainable agricultural systems [...] and establishes a dialogue between traditional

2 "STAV" stands for *Sciences et technologies de l'agronomie et du vivant* (Agronomy and Living Sciences and Technologies).

agricultural practices used for hundreds of years by peasants and modern knowledge on agronomy, ecology, and the social sciences". Here the appearance of two concepts drawn from the critical dictionary is evident: the notions of systemic analysis and of complexity. In fact, in the high school agricultural education curricula, "the aim is to get learners to reflect on agricultural systems and practices according to various goals and contexts, taking ecosystems into account, in an environment marked by changes and uncertainty, particularly climatic changes" (see Ministry of Agriculture 2019). Here other sustainable development skills appear, such as foresight and social responsibility. In other words, there are two designations: agro-ecology (with a hyphen) can use glyphosate and other phytopharmaceuticals, whereas agroecology will do without any chemicals. The values carried by one and the other are sometimes very different.

We postulate that these different elements contribute to agroecological literacy.

6.3. Technical knowledge and political movements

Agroecology is often considered as a body of scientific knowledge combining agronomy and ecology, as a peasant[3] political movement and as a set of agricultural practices and techniques. The term emerged in the scientific literature in the early 1980s. The political dimension of agroecology induces numerous practices and values, and there are many who apply them. According to Catalogna (2018), "the intensification of ecological processes proposed by agroecology is highly dependent on the agricultural context. A renewal of knowledge on farming is necessary to adapt each agroecological practice". Nevertheless, there is a common ground: reducing erosion and cultivating the soil, increasing soil fertility, enhancing functional biodiversity, preserving water resources, promoting genetic diversity, increasing the food autonomy of herds and valuing co-products.

According to Reboud and Hainzelin (2017), the "quest for a break with 'classic' agronomy questions the ability of interdisciplinarity to provoke (and then to justify?) an evolution of perspectives via the concepts of sobriety, efficiency, resilience or robustness".

3 In Latin America in the 1970s, there was a social movement development – generally inspired by left-wing trade unions – reuniting peasants who defended access to land, rural development and agriculture with respect for the environment. More or less politicized, they claimed food autonomy and sovereignty for peasant populations. They stimulated the development of agroecological "practices" and their transmission from peasant to peasant (*Campesino a Campesino*). *La Vía Campesina*, an autonomous and independent movement, defended sustainable small-scale agriculture as a means for promoting social justice and the dignity of populations. It defended food sovereignty and small agro-ecological agriculture.

The founding fathers of agroecology, Gliessman (2015) and Altieri (2018), provide complementary visions of agroecology. Altieri presents agroecology as the science of natural resource management and defines it according to five principles. According to Gliessman, agroecology is the application of ecology to the study, design and management of sustainable agrosystems.

Here, the context of agroecological literacy on how to "produce differently" is approached from the perspective of the contributions of ancestral knowledge. This can sometimes be idealized or refuted, which explains the importance of teaching agroecology, of its politicization. According to Arrignon (2020), we should even speak of the ministerial depoliticization of this agricultural approach, the alleged difficulty of teaching it, and the need for it in the current climate change context, with the "eating better" agricultural policies being called upon to revise their approaches regarding the protection and sustainability of natural resources.

In addition, for Arrignon (2020), "two forms of depoliticization seem to be at work in this reform [agroecological transition and sustainable development], one involving the negation of conflicts within the agricultural worlds, and the other implying the disengagement of the central State". In that sense, there is a "redefinition of challenges".

In Arrignon's words, "the desire not to stigmatize agriculturalists on the basis of their practices is found in the claim not to oppose the economic issue of production to the ecological issue of environmental protection".

Political questions arise once again (what agricultural potential in agroecology?) together with the values around agro-environmental policies in the fight against global warming and its agricultural consequences. The latter risk impacts crops (production, among other aspects) and, therefore, the content of dishes, both in qualitative and in quantitative terms. "Eating better" is one of the major agricultural challenges of the 21st century.

6.4. Knowledge conflicts and conflicts of values: the question of direction in the circulation of knowledge

If "the rationality of science and technology is positivist" (Fabre 2018), then we can wonder about society's view on agroecology. In this sense, agroecology is a politicized discipline born from social ("revolutionary") movements, which seems to fit badly with the republican demand for a neutral school devoted to the teaching of "truth", in line with the universal reason developed by the Enlightenment philosophers (Audigier et al. 2015): "As if the (didactic) ethics of an education in ecological citizenship were not reconcilable with impartial communication" (Urgelli 2009).

Agricultural literacy would be closely related to agronomic curricular guidelines and scientific literacy, emphasizing the scientific research production process, enhancing problem solving skills and acknowledging a major societal dimension. In agricultural high schools, agroecology is a discipline taught in a variable way, depending on the institutions' choices and policies, the place and strategies for agricultural exploitation and the knowledge and values promoted by the teaching team. The teaching of agroecology is therefore complex.

We postulate that these scientific and technical data adapted to the territory, respecting both biotic and abiotic natural resources, should enable diverse learners to interact with an agricultural world not only shaped by science and technology (standardized scholarly knowledge) but also by foundations (knowledge from experience) – soil and climate – without forgetting the vagaries of territory – weather, the passage of pests. These data are presented in the form of curricular markers for learners or for people in professional retraining who aim to "produce differently" and who are anchored in a specific territory, rather than disconnected from the ground.

Agroecological practices are partly based on traditional knowledge drawn from peasants[4] in tropical and subtropical countries, in particular. In no way is this knowledge opposed to a scientific approach: studying the peasants' agroecological practices should make it possible for scientific agroecology to understand and improve their conditions for success.

However, by highlighting "profane" knowledge, the relationship between scientists and agriculturalists is modified, the latter being necessarily more involved in research programs (design, development); this is "participatory" research. We might wonder whether researchers can refute field experience.

Teaching how to "produce differently" is one of the axes of agroecological literacy. It necessarily has to rely on the field to learn from it so that this discipline does not become like certain productions: disconnected from the ground.

6.5. Towards coherent criteria for analyzing agroecological literacy

6.5.1. *Essential curriculum elements*

As far as reference sources are concerned, contributing knowledge can be easily identified: biological and earth sciences (ecology, pedology, crop science,

4 In the list of "land occupations", the denomination "peasant" does not appear and is therefore not officially recognized. The profession is "agriculturalist". The term "agro-peasant" has no legitimate value, except in the eyes of those who practice agroecology daily.

zootechnics, etc.), economics (legislation, management, etc.) and transversal knowledge. On the other hand, constitutive practices are more complex to define, as are the political prescriptions with their injunctive values.

According to our previous argument, constitutive practices must be bottom-up and engage the agricultural approach; the roles of agriculturalists (here, agro-peasants) would be production management, equipment maintenance, company accounting and those related to the care taken in production and marketing, while political values and injunctions must derive from the political components that gave birth to agroecology ("better eating", the fight against climate change and the loss of biodiversity, resource conservation, etc.); in short, the "how?" is still to be defined.

On top of everything else, the curricular dimension related to sustainability (systemics and complexity) leads us to hypothesize that the links between the different elements constitute coherent agroecological literacy.

Following this reasoning, we produced the model to be used in our case study (Figure 6.2).

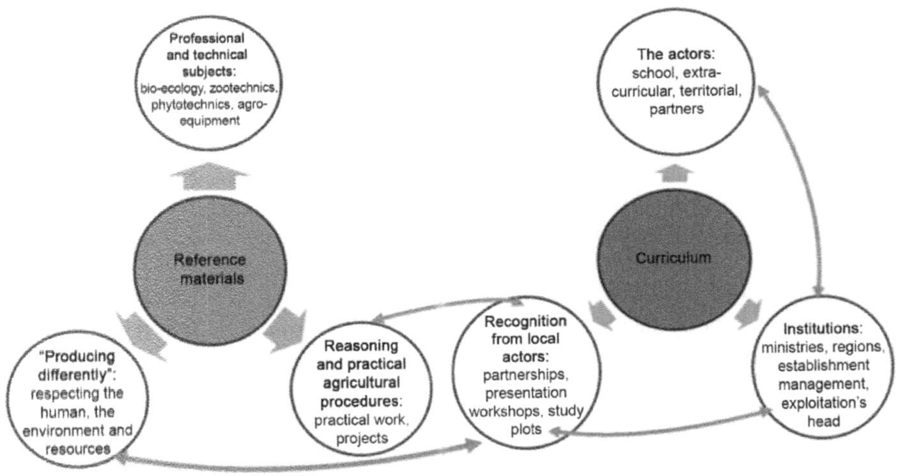

Figure 6.2. *Necessary links to a curriculum coherent with agroecology in agricultural education*

6.5.2. *A practical method for measuring necessary curricular links*

In addition to the model in Figure 6.1, Figure 6.2 shows double circulation arrows between the different poles. These privileged ways of circulation of knowledge, experiences and practices can contribute curricular consistency and reference materials to balance the expectations (e.g. reduction in the use of inputs) and needs (e.g. means of communication) from each pole. These will be more resilient in the face of societal demands, so agricultural education will be more adapted to the territory's (top-down and bottom-up) demands. Institutions collaborating with the local actors will be able to work hand in hand "to produce differently", as territorial specificities will be taken into account.

Our research hypothesis is as follows: to arrive at a coherent curriculum, and in accordance with our argument, several dimensions must then be taken into account (Figure 6.2).

At the same time, if the curriculum must respond to a systemic approach, links must exist between the different elements of the model. To measure it, we used a Likert scale (from 1 to 5, with a double classifying direction) showing that there are bottom-up (going from right to left) and top-down (from left to right) approaches which can meet (Figure 6.3). In other words, the aim was to measure whether the curriculum development process includes the links that contribute to its coherence by means of field surveys. The absence of links would signal an obstacle (highlighted by a blank between the two parts in Figure 6.3).

A survey makes it possible to visualize the links and interweaving between actors, which makes it possible to measure the coherence of the resulting curriculum. This work carried out in the field and in class enables a hybridization of knowledge, with a dual approach towards the notion studied (top-down, in our case represented by institutions, and bottom-up, represented by agro-peasants and teachers).

Regarding the criteria followed, it is necessary to add a cursor taking into account the top-down (right to left) and the bottom-up (right to left) approaches.

We find it particularly relevant to be able to determine the influence of the local approach (bottom-up) and the more institutional approach (top-down) for each pole (Figure 6.3), which can represent the commitment of each actor (institutional or local) to agroecology. If an institutional demand is strong towards local actors (e.g. reducing GHGs), it should be awarded a score of 5 points in agreement with the local actor (institutional injunction).

If the institution's demand disagrees with the local actor (multifactorial causes: lack of time, lack of technical means, in disagreement with local agriculture, etc.), the latter's commitment should only be awarded one point (no/too little support).

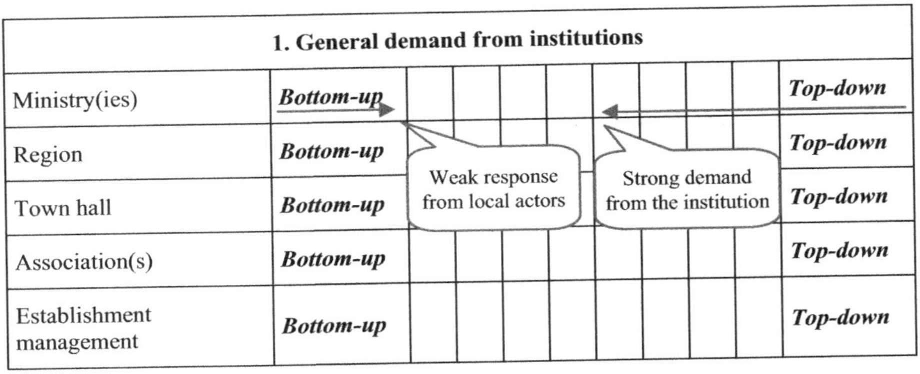

Figure 6.3. *Actor demand grid around the notion of agroecology*

If a significant distance must be covered between the two poles, this reflects a lack of consistency between the demand and the receiver (without prejudging one to be more correct than the other). It is also important to consider that the misunderstanding could have originated in the local and that the institution did not take it into account.

Thus, we can provide an indicative score in order to understand whether the dynamism originates in the local approach (agro-peasants, students, etc.) or in the institutions (school directors, communities, etc.).

If the two poles come together, then we can say a bridge is being formed and that local actors and institutions are coming together around the project(s) proposed and to be implemented; a certain coherence is established between the institution and the local actor. Through this model, we can deduce the barriers and levers tending towards "producing differently" by teaching agroecology in relation to territorial contexts.

Having presented the curricular challenges and the methodology used, our case study helps us answer the questions on the local coherence of an agroecological curriculum, as well as regarding the links and roles of territorial anchorages.

6.6. Case study

In the case study, we sought to determine whether the various criteria presented above had achieved curricular coherence in the case of an agricultural high school that benefits from the support of a biosphere reserve territory.

To carry out this work, we described the agricultural activity of agro-peasants in the Mont Ventoux biosphere reserve in relation to the Carpentras *Lycée d'enseignement général, technologique et/ou professionnel agricole* (LEGTA) producer store and farm visits. During these visits, we were able to complete a questionnaire (to contextualize the agricultural exploitation) and an interview grid (to draw a portrait of the agro-peasants in terms of their "relationships"). Thanks to these data, it was possible to determine the barriers and levers for each actor.

6.6.1. *Biosphere reserves, levers to "produce differently"?*

Biosphere reserves obey "common criteria" defined by UNESCO in its Sustainable Development Goals roadmap. UNESCO defines them as a special place to test the interdisciplinary approach to understanding and managing changes and interactions between social and ecological systems, including conflict prevention and biodiversity management (Chitanava 2018). Biosphere reserves potentially become territories compatible with "producing differently" and the associated agroecological values; they are like places of appropriation of environmental challenges by the local actors. The conservation management of these spaces by the population makes them places for research and experimentation and a springboard for alternative production, as in the Cévennes, for example, where they prompted the revival of the chestnut grove (Crosnier 2006). There are often many alternative food networks, certifications and labels for local products and short circuits between producers and consumers. These networks are seen as promoting value chains in rural areas (Olah and Zigrai 2004). They have created initiatives like the Dordogne basin biosphere reserve, which instituted the "Food and Agriculture" trophy, in order to favor the use of local products and those promoting regional identity.

In this context, agroecology holds a special place, which goes a step beyond "producing differently"; it envisages another model of production (with respect for the environment, for the people), of distribution (short circuit) and of resource consumption (both abiotic resources and food). It makes part of a complex societal approach to sustainability, which is the last point we focus on.

6.6.2. *The Mont Ventoux biosphere reserve and the agricultural school of Carpentras*

Here, we focus on the agricultural school of Carpentras, linked to the Mont Ventoux (84) biosphere reserve (BR). This reserve has made it its goal to teach the principles of agroecology on a territory mainly occupied by vines, which are fairly large consumers of phytopharmaceutical products. Under the guidelines of its operations manager, the agricultural exploitation attached to this LEGTA embarked on local agricultural reflection beyond the idea of "producing differently". On the national level, it is one of the 30 agricultural high school farms out of 192 to have made this decision.

Within a territory such as the Ventoux biosphere reserve, where agroecology is driven by the dynamism of agro-peasants and the possibilities offered by such a place, we can hypothesize that the LEGTA's teaching is coherent. Let us see if this is the case.

Using the model described below, we also explored the work and reflections of the Carpentras LEGTA in its links to agroecology and the Mont Ventoux biosphere reserve and conducted interviews with the operations manager, the head of experimentation and a teacher in technical matters.

Data collection made it possible to visualize the interweaving and connections that the agro-peasant establishes with other local actors within the biosphere reserve.

This research enabled us to draw a portrait of agro-peasants in the Mont Ventoux biosphere reserve, as well as to better understand the state of mind in which the teaching of agroecology takes place within the LEGTA.

1. General Demand from Institutions										
Agriculture and Ecology Ministry	*Bottom-up*									*Top-down*
PACA region	*Bottom-up*									*Top-down*
Carpentras city hall	*Bottom-up*									*Top-down*
Association(s)	*Bottom-up*									*Top-down*
Establishment management (LEGTA, CFPPA, agricultural exploitation)	*Bottom-up*									*Top-down*
2. Demand for Agroecological Teaching										
Team project	*Bottom-up*									*Top-down*
Teacher's group project	*Bottom-up*									*Top-down*
Teacher's individual project	*Bottom-up*									*Top-down*
Students' group project	*Bottom-up*									*Top-down*
Student's individual project	*Bottom-up*									*Top-down*
3. Agro-peasant Demands										
Projects with the institutions	*Bottom-up*									*Top-down*
Projects with high school exploitation	*Bottom-up*									*Top-down*
Projects with the teachers	*Bottom-up*									*Top-down*
Projects with the students	*Bottom-up*									*Top-down*

Figure 6.4. *Grid showing actors' commitment to agroecology for the Carpentras LEGTA. For a color version of this figure, see www.iste.co.uk/barthes/biosphere1.zip*

6.6.3. *Case study results*

The results of the field surveys have been broken down into two parts.

For the first result, we took into account the relationships that bring together, or do not, the top-down (institutions) and the bottom-up (agro-peasants and education) approaches through three ideas: the general demand from institutions in relation to agroecology (curriculum), the demand for the teaching of agroecology (articulation within the teaching team, link with the students, prescribed curriculum) and the demand from agro-peasants (projects that may or may not be articulated around the general and technological high school).

Figure 6.4 shows that the demand to practice/teach/promote institutional agroecology is strong. The actors in the field (agro-peasants and/or teachers) do not seem to be involved (lack of knowledge, lack of networking, etc.). Moreover, the institutions encourage the LEGTA team to implement projects that are not always feasible (gap between the demand and the possibilities offered by the field), and there is a marked match between the actors in the field and the LEGTA.

Our study highlights a disparity between field demands and institutions.

As far as local investment (projects) in the LEGTA is concerned, this acts more as a driving force and goes further than what the institutions expect. The link between the two poles (field and high school) is established by the training reference materials of the different high school sections (in particular, via the technical skills to be acquired).

It is also patent that agro-peasants have difficulty finding a place within the LEGTA, probably due to a lack of time in this case.

For the second result, we took into account what happens within the Carpentras LEGTA and the existing relationships between various poles that seem necessary for the success of coherent in situ learning of agroecology.

Figure 6.5 shows that there are links between almost all the actors (institutions at various scales, agro-peasants and those within the LEGTA). This provides evidence of a wealth of exchanges between the actors and the local potential of agroecology, which seems to be favored by the biosphere reserve setting, similar to an ecosystem. The more interactions between the actors (institutions, teaching staff, establishment management and agro-peasants), the richer the overall system and the more resilient it is in the face of disturbances.

In the other LEGTAs, our studies show that this model is more rudimentary (no place for sale and no visibility of the farming work, for example).

Unlike many other high schools studied (30 out of 191 LEGTA institutions are engaged in reflection and completely reviewing their exploitation system (Ministry of Agriculture 2019)), the Carpentras high school genuinely encourages a connection between values, field knowledge and scholarly knowledge. Thanks to the relay role played by agricultural exploitation, there is an actual knowledge exchange: the circulation of knowledge is not unidirectional (Figure 6.5).

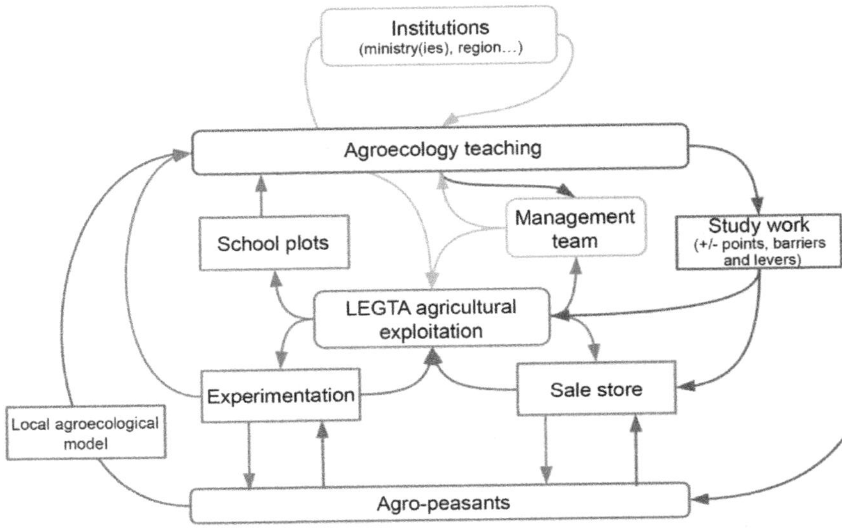

Figure 6.5. *Multiple circulation of top-down and bottom-up agroecological knowledge components – the case of Carpentras LEGTA. For a color version of this figure, see www.iste.co.uk/barthes/biosphere1.zip*

6.7. Discussion

6.7.1. *What curricular coherence should exist for teaching to "produce differently" in the biosphere reserve?*

The Ministry of Agriculture, by means of the LEGTAs, could have gone further in implementing teaching about "how to produce differently" via agroecology. If they had the will, would the proximity of an exceptional territory (biosphere reserve, parks, etc.) enable operating managers to take the plunge more easily, the structures and dynamism being perhaps more significant than elsewhere?

We can therefore understand that the teaching of agroecology as practiced by several disciplines still raises questions on how to teach it.

By defining their location, biosphere reserves are conducive to simplified access to agroecology; they facilitate experimentation in practical agroecology, they prompt the preservation of the environment and popularize the notion of how to "produce differently" for a vaster audience. On the other hand, they are not yet visible enough for most people, which acts as a barrier, seriously counterbalancing all of their benefits.

There seems to be a lack of organization of agroecological exploitation networks in the Mont Ventoux biosphere reserve, not for a lack of interest or energy but due to an opaque screen maintained by conventional agriculture (particularly in viticulture and arboriculture).

6.7.2. *The teacher's posture in relation to agroecology*

In agroecology, we can regret the relative absence of theoretical articulation between values and knowledge: knowledge coming from agro-peasants in the field clashes with scholarly knowledge, and knowledge exchanges are only established unidirectionally, "descending" from scholarly knowledge to profane knowledge.

Could agroecology still be a reservoir of controversial knowledge in its learning and knowledge dimensions?

The teaching of agroecology collides with several agricultural methods, such as extensive/intensive reasoned/biological methods, and the teacher's commitment to neutrality seems undermined in the face of products resulting from those types of agriculture later found on our dishes. Still, while values are important in "education for", their teaching is not self-evident and carries the risk of activating latent value conflicts.

The teaching team is not yet unanimous on the values to be taught around the notion of agroecology; despite having received unequal teachings, the students gain access to a complete version of the challenges associated with this notion.

> Telling the truth calls into question the bond between the one who speaks the truth and the one to whom this truth is addressed, and this takes courage (Lange and Kebaïli 2019).

According to what the students reported within the LEGTA, agroecology seems obvious to them in the technological branches. This is less frequent in professional branches, where the team's view of agroecology is not yet unanimous regarding how it should be taught.

6.7.3. *The role of the territory in the implementation of a consistent curriculum*

In the case of agroecology, the agro-peasant works, the agronomist validates and conceptualizes it and then the agro-peasant uses this validated knowledge… without their field knowledge ultimately gaining a central place in the agronomist's thinking.

For the Biovallée (Drôme) and its representations exhibited during the forum "Origin, Development and Territory", the soil of the Ventoux territory – with its peasants practicing agroecology and the joining of the movement on how to "produce differently" – was a motivation and a driving force seemingly gaining ground despite everything and penetrating the walls of the "neutral" teaching of this discipline.

In Figure 6.5, we can observe a set of actors, local actors or those closest to the school establishment, with a highly decisive political influence compared to distant actors.

The operations manager has a decisive role in the impact of agroecology within this LEGTA, supported by their team, in particular its head of experimentation, and by the local agro-peasant dynamism of the Mont Ventoux biosphere reserve.

Here, a variety of projects carried out by pupils/trainees/apprentices force the professionals of the earth to push the door of this store and to go and watch the productions of the school's exploitation. This showcase is a successful and tangible proof that all the efforts in methodology and reflection and different practices can lead to healthy and profitable products agro-peasants are proud of. In other territories without this biosphere reserve feature and the energy provided by the agro-peasants (known within the school via the producer store), agroecology is taught but has no particular echo outside the classroom.

Without this double support (biosphere reserve, exploitation philosophy), this teaching would be drowned and become entangled in the landscape of different agricultural practices.

6.8. Conclusion

Agroecology is at a crossroads: biosphere reserves, open territories for reflection, would be a favorable ground for the emancipation of this multifactorial knowledge trapped between scholarly and profane knowledge and the politicized dimension it contains.

Despite all, in this territory, the LEGTA contributes to the dissemination of this agricultural ideology and this discipline; it favors an evolution in mentalities, either through the appropriation and acceptance of this subject by listening students or at least its introduction to more reluctant students.

The Carpentras LEGTA will have written its own agroecological literacy thanks to the local wishes relayed by the management team and teachers, the operations manager and the head of experimentation and the agro-peasants from the fertile territory of Mont Ventoux biosphere reserve.

We may wonder whether this model is transposable and if the tools presented could make it possible to make an inventory of other establishments wishing to teach how "to produce differently".

6.9. References

Akitsu, Y., Ishihara, K.N., Okumura, H., Yamasue, E. (2017). Investigating energy literacy and its structural model for lower secondary students in Japan. *Science Education*, 12(5), 1067–1095.

Albe, V. (2011). Finalités socio-éducatives de la culture scientifique. *Recherches en éducation*, 174, 119–138.

Altieri, M.A. (2018). *Agroecology: The Science of Sustainable Agriculture*, 2nd edition. CRC Press, Boca Raton.

Altieri, M.A. and Nicholls, C.I. (2004). An agroecological basis for designing diversified cropping systems in the tropics. *Journal of Crop Improvement*, 11(2), 81–108.

Arrignon, M. (2020). La transition agroécologique : une politique de développement durable comme les autres ? *VertigO*, 20(1).

Audigier, F. (2012). Les Éducations à... et la formation au monde social. *Recherches en didactiques*, 14(2), 47–63.

Audigier, F., Crahay, M., Dolz, J. (2006). *Curriculum, enseignement et pilotage*. De Boeck Supérieur, Brussels.

Audigier, F., Sgard, A., Tutiaux-Guillon, N. (2015). *Compétences et contenus ; les curriculums en questions*. De Boeck Supérieur, Brussels.

Azevedo, Z. and Marques, M. (2017). Climate literacy: A systematic review and model integration. *International Journal of Global Warming*, 12(3/4), 414.

Barthes, A. and Alpe, Y. (2012). Les "éducations à", un changement de logique éducative ? L'exemple de l'éducation au développement durable à l'université. *Spirale*, 50(1), 197–209.

Barthes, A. and Alpe, Y. (2018). Les "éducations à", une remise en cause de la forme scolaire ? *Carrefours de l'éducation*, 45(1), 23–37.

Barthes, A., Romagny, B., Aderghal, M., Lange, J.M. (2021). Enjeux politiques de l'internationalisation des curricula de l'enseignement supérieur. Regards croisés autour du programme Erasmus + pour les réserves de biosphères.

Bianchi, F.J.J.A., Booij, C.J.H., Tscharntke, T. (2006). Sustainable pest regulation in agricultural landscapes: A review on landscape composition, biodiversity and natural pest control. *Proceedings of the Royal Society B: Biological Sciences*, 273(1595), 1715–1727.

Catalogna, M. (2018). Expérimentations de pratiques agroécologiques réalisées par des agriculteurs : proposition d'un cadre d'analyse à partir du cas des grandes cultures et du maraîchage diversifié dans le département de la Drôme. PhD thesis, Avignon University, Avignon.

Chitanava, M. (2018). Biosphere reserve. The role model of regions for sustainable development. *ISET Economist Blog* [Online]. Available at: https://iset-pi.ge/en/blog/154-biosphere-reserve-the-role-model-of-regions-for-sustainable-development.

Cibien, C. (2006). Les réserves de biosphère : des lieux de collaboration entre chercheurs et gestionnaires en faveur de la biodiversité. *Natures Sciences Sociétés*, 14(1), 84–90.

Clavier, L. (2018). Processus d'institutionnalisation et éducation : l'enseignant médiateur des transformations de l'élève. Éducation et socialisation. *Les Cahiers du CERFEE*, 48.

Coquidé, M. (2018). Ignorance, hasard, incertitude, risque : des enjeux pour une éducation scientifique émancipatrice. *Éducation scientifique et technologique et émancipation*, 34.

Crosnier, C. (2006). Biodiversité et pertinence des pratiques locales dans la réserve de biosphère des Cévennes. *Revue internationale des sciences sociales*, 187(1), 159–168.

Fabre, M. (2018). Savoir et valeur. Pour une conception émancipatrice des "Éducations à". Éducation et socialisation. *Les Cahiers du CERFEE*, 48.

Forquin, J.-C. (2008). *Sociologie du curriculum*. Presses universitaires de Rennes, Rennes.

Gibert, A.F. (2020). Éduquer à l'urgence climatique. *IFE*, 44.

Girard, N. (2014). Gérer les connaissances pour tenir compte des nouveaux enjeux industriels : l'exemple de la transition écologique des systèmes agricoles. *Revue internationale de psychologie*, 19(1), 51.

Gliessman, S.R. (2015). Agroecology: A growing field. *Agroecology and Sustainable Food Systems*, 39, 1–2.

de Ketele, J.M. (2010). La pédagogie universitaire : un courant en plein développement. *Revue française de pédagogie. Recherches en éducation*, 172, 5–13.

Landel, P.A. and Girard, S. (2020). Interroger la (possibilité) de transition (agro)écologique territorial. Le cas de la vallée de la Drôme-Biovallée [Online]. Available at: http://www.psdr.fr/archives/INS1059PDFN1.pdf [Accessed 13 October 2022].

Landel, P.A. and Senil, N. (2009). Patrimoine et territoire, les nouvelles ressources du développement. Développement durable et territoires. *Économie, géographie, politique, droit, sociologie*, 12.

Lange, J.M. and Kebaïli, S. (2019). Penser l'éducation au temps de l'anthropocène : conditions de possibilités d'une culture de l'engagement. *Éducation et socialisation. Les Cahiers du CERFEE*, 51.

Lebeaume, J. (2019). Précisions sur la "forme curriculaire" et distinction entre pratiques constitutives et savoirs contributifs. *Éducation didactique*, 13(1), 43–59.

Lee, L.S., Lee, Y.F., Altschuld, J.W., Pan, Y.J. (2015). Energy literacy: Evaluating knowledge, affect, and behavior of students in Taiwan. *Energy Policy*, 76, 98–106.

Legardez, A. (2017). Propositions pour une modélisation des processus de didactisation sur des questions socialement vives. *Socially Acute Questions*, 5(2), 79–99.

Leménager, T. and Ehrenstein, V. (2016). Des principes agroécologiques à leur mise en pratique. *Revue Tiers Monde*, 226(227), 3, 65–93.

Matagne, K., Hoedt, A., Piron, J. (2017). La Faute de l'orthographe. *Textuel*, 1, 148.

Ministry of Agriculture (2014). Environnement : le changement climatique, une réalité qui s'impose à tous. Report, Bureau de la stratégie environnementale et du changement climatique, Sous-direction de la biomasse et de l'environnement.

Ministry of Agriculture (2019). Bilan 2014–2018 du plan "Enseigner à produire autrement". Report, Ministry of Agriculture.

Nédélec, L. (2018). L'éducation aux sciences dans un monde incertain : comment les enseignantes appréhendent-elles les incertitudes de la question vive de la transition agroécologique ? Étude de cas auprès d'enseignantes de l'éducation nationale et de l'enseignement agricole en contextes de formation. PhD thesis, University of Toulouse-Jean Jaurès, Toulouse.

Neumann, K., Viering, T., Boone, W.J., Fisher, H. (2013). Towards a learning progression of energy. *Journal of Research in Science Teaching*, 50(2), 162–188.

Olah, B. and Zigrai, F. (2004). The meaning of the time-spatial transformation of the landscape for its sustainable use (a case study of the transition zone of the Pol'ana Biosphere Reserve). *Ekológia*, 23, 231–243.

Pestre, D. (2006). *Introduction aux Science Studies*. La Découverte, Paris.

Reboud, X. and Hainzelin, É. (2017). L'agroécologie : une discipline aux confins de la science et du politique. *Natures Sciences Sociétés*, 4, 64–71.

Roth, X., Barthes, A., Cohen, J. (2020). Plaidoyer pour les antivaleurs ? *Penser l'éducation*, 47, 89–108.

Stassart, P.M., Baret, P., Grégoire, J.-C., Hance, T., Mormont, M., Reheul, D., Stilmant, D., Vanloqueren, G., Visser, M. (eds) (2011). *Agroécologie*. Educagri, Dijon.

Tscharntke, T., Tylianakis, J.M., Rang, T.A., Didham, R.K., Fahrig, L., Batáry, P., Bengtsson, J., Clough, Y., Crist, T.O., Dormann, C.F. et al. (2012). Landscape moderation of biodiversity patterns and processes – Eight hypotheses. *Biological Reviews of the Cambridge Philosophical Society*, 87(3), 661–685.

UNESCO (2014). Feuille de route pour la mise en œuvre du Programme d'action global pour l'éducation en vue du développement durable. Report, UNESCO.

Urgelli, B. (2009). Logiques d'engagement d'enseignants face à une question socio scientifique médiatisée : le cas du réchauffement climatique. PhD thesis, University of Lyon, Lyon.

7

The Sustainable Management of Biosphere Reserves: What Are the Challenges for Agricultural Education?

7.1. Introduction

As early as the 1940s, the United Nations Educational, Scientific and Cultural Organization (UNESCO) took an interest in environmental issues, mainly addressing the question of nature conservation. The beginning of those years was marked by the growing importance of environmental considerations and the emergence of reflection on the biosphere at an international level. These new concerns were materialized by the creation of the United Nations Environment Programme in December 1972 and the launch of the Man and the Biosphere Programme by UNESCO. At present, the guidelines promoted by international policies aim to support the agroecological and social transition, as well as to reinforce the Sustainable Development Goals by 2030. The main missions of the Man and the Biosphere Programme are: (1) to preserve natural and cultural diversity, (2) to build space management models and places for experimenting with sustainable development and (3) to constitute privileged places for research, continuous monitoring of the environment, education and training.

As a training organization preparing students for careers related to nature and the living while providing training for farmers and actors in the rural sphere, agricultural education is targeted by this program, which also aims at education and training. After 1985, agricultural education undertook a major renovation of all diplomas, striving to meet the significant societal demand in favor of the environment. In 2014, the "Agroecology for France" project led the Ministry of Agriculture and

---

Chapter written by Nina ASLOUM, Guillaume GILLET and Laurent BEDOUSSAC.

Food to include food and forests in its preparatory work for the upcoming law on agriculture (Law No. 2014-1170), a national strategic project for agricultural education called "teaching to produce differently". The project aims to encourage agriculture and agricultural education to change their practices so as to develop alternative production systems, which, among other things, should enable better management and preservation of rural, urban and natural spaces and biosphere reserves. The guidelines state that "the transition towards new, more sustainable production systems is based on a major change in the frameworks of thought and knowledge acquisition modes and practices". Teachers are called upon to play a decisive role in implementing these new orientations at the origin of a major reform of the agricultural training system in order to acknowledge the ecological, pedagogical and educational dimensions of the agroecological transition.

However, these directives, which reinforce those related to sustainable development, are not self-evident because the agroecological transition is based on a certain number of socio-epistemic uncertainties which still need to be called into question (epistemic uncertainties, the effects of practices, the actions to be taken, the actors involved and uncertainty in itself, etc.). Teachers facing these changes express difficulties in their ability to take a stand regarding two key agroecology-related disciplines: agronomy and ecology (Stassart et al. 2012). These difficulties are exacerbated by the abundantly plural reference knowledge (academic, professional, social, etc.) covered by agroecology, which relies on knowledge most often still under construction or even controversial (emerging knowledge, sometimes based on beliefs). While teachers are used to summoning relatively stable knowledge, they sometimes have to deal with unstable concepts that have a complex nature and interact with their own uncertainties and those of their students, in a context where prescriptions are evasive, even discretionary (Gillet et al. 2017). These prescriptions – left to the teacher's criterion as to the notion of "good agricultural practice" – leave room for a significant degree of subjectivity, depending on the common sphere of belonging invoked (Béguin et al. 2011) and the reference agricultural models used.

In this chapter, we first present the emergence of agricultural development prompted by agroecology. Then, from the directives assigned to agricultural education, we strive to analyze the perceptions that teachers have of "producing differently", using the social representations theoretical framework as a stepping stone (Abric 1994). This prerequisite seems necessary in order to understand the meaning that teachers give to agroecology and to the expression "to produce differently", as well as its associated techniques and practices. In addition, this approach allows us to propose training tips for teachers to support them in an enlightened posture vis-à-vis the agroecological transition.

7.2. Agroecology, from its emergence to the change of agricultural model

7.2.1. *Agroecology: an ever-evolving, polysemous concept*

Originating in Latin America, agroecology had an impact on the social movements that developed in France around the 1980s. By advocating the acknowledgment of social inequalities, agroecology reinforces sustainable development goals. Making practices and standards more respectful of the environment and the management of spaces engages important social, environmental and political goals. Spanning its emergence and its circulation in the different arenas (social, professional and scientific), the agroecological transformation is endowed with a polymorphic character whose definition has evolved over time. For Wezel et al. (2009), agroecology is an alternative to conventional agriculture and simultaneously denotes a set of agricultural practices, a science and a social movement. Altieri (1987) and Gliessman (1990) consider agroecology as a set of methods and practices for rethinking the links between agriculture and ecosystems to guarantee the preservation of resources. Gliessman (1990) emphasizes that agroecological production methods are based on the use of principles and concepts derived from ecology and inspired by biodiversity and its associated natural processes (the cycle of the elements and biological balance between organisms).

Tomich et al. (2011) add that agroecology is "a disciplinary ensemble fed by the crossing of agronomic sciences (agronomy, zootechnics), ecology applied to agroecosystems and human and social sciences (sociology, economics, geography)". However, agroecology can be considered in a broader sense that is not limited to agricultural production and the scale of the agrosystem. In this broader vision, agroecology integrates the entire food system and aims at a transition towards viable systems with respect for people and their environment, requiring the support of transdisciplinary and multi-actor approaches (professionals in the agricultural sector, scientists, public policies and social movement actors). Francis et al. (2003) propose to consider agroecology as "the integrative study of the ecology of the whole food system, acknowledging the ecological, economic and social dimensions". As Stassart et al. (2012) have pointed out, these different definitions show "that there is no single way to define and work on agroecology".

Ultimately, agroecology is a scientific discipline, a social movement and a set of agricultural practices whose multiple dimensions are, as Wezel et al. (2009) have emphasized, situated in time, since "the application of the term is highly dependent on the development of history and epistemology, which provide the basis, scope and validity for the use of the term in different countries". Stassart et al. (2012) also add that agroecology "cannot be considered without taking into account the relationship between science and society". Although defined and recognized as a discipline or interdiscipline, the reference frameworks and alternative practices invoked by agroecology are, as various authors have pointed out, yet to be clarified scientifically and rooted in reference practices in a constantly changing context, particularly due to climate change.

7.2.2. *Agroecology for a real change in the agricultural model?*

The "Agroecology for France" project was initiated in 2014 by the Ministry of Agriculture and Food to break with productivist agricultural practices which had reached their limit (loss of biodiversity; soil erosion; air, water and soil pollution; impact of the carbon footprint on global warming, etc.). The guidelines thus state that "the transition towards new, more sustainable production systems should be based on a major change in the frameworks of thought and knowledge acquisition modes and practices".

The public policy challenge was therefore to promote agroecology as a new model capable of ensuring food self-sufficiency through agricultural production, while meeting demands for quality and health safety. However, for Bellon and Doré (2019), the implementation of agroecology questions the current organization of agri-food systems in terms of sustainability and calls for reconsidering the links between agriculture, environment, science, society and politics. This transition requires the implementation of a deconstruction and recomposition process in different domains (social, technological, political and cognitive), which can lead to relative confusion.

This new framework of thought challenges us and, like Bellon and Doré (2019), we can wonder whether it actually reflects a "deeper current with multiform paths which could build new agricultural and food systems better integrated in society". In this sense, Bellon and Ollivier (2012) encourage us to be vigilant.

Despite the unquestionable conceptual innovation (intensified in knowledge and in ecosystemic processes), we wonder whether – intentionally or not – the vagueness in the use of these terms (agroecology, agriculture conservation, etc.) involves a certain pragmatic consensus around the lowest common denominator and is restricted to the above-mentioned few forms of agriculture.

In addition, this new political injunction cannot be sufficient on its own and must necessarily go through a process of appropriation and adhesion because, as Abric

(1994) has cleverly pointed out: "It is not enough for an individual to be engaged in a practice for them to recognize it as their own and to appropriate it. It is still necessary for the practice to appear to them as acceptable in relation to the system of values they cherish."

7.3. Social representations

7.3.1. *The theoretical framework of social representations*

For Moscovici (1984), the social representation of individuals is:

> A way of interpreting the world and thinking about daily reality, a form of social knowledge that the person constructs more or less consciously from what he/she is, has been and projects and which guides behavior. Social representation is the mental activity deployed by individuals and groups to fix their positions in relation to situations, events, objects and communications that concern them.

Jodelet (1997) specifies that representations are understood as systems of interpretation that govern "our relationship with the world and others, guide us and organize behavior and social communications". Then, the question comes down to understanding the social or behavioral practices of a group, going through an in-depth analysis of the object's representations by the groups. In short, a social representation is made up of an (organized and structured) ensemble of information, opinions and beliefs and is invariably constructed within a group and in relation to an object, and as Jodelet (1997) has stated: "There is no representation without an object."

These elements raise the question of the transition from the individual to the group and how the meaning of an object structures the judgments and perceptions of individuals and prompts them to act (representations in and for the action).

7.3.2. *The structural approach to social representations*

Seeking to understand social representations of teachers about how to produce differently, we rely on the theory of the central core and the peripheral elements (Flament 1981; Abric 1994; Vergès 1994). The reference to this theory seems appropriate to interpret the complexity of the social object studied herein. In addition, as Garnier and Sauvé (1999) have emphasized, relying on "the analysis of social representations can prove promising for this purpose, especially since it can be part of a reflexive perspective with a view to contributing to the transformation of social practices".

The theory of the central core proposed by Abric (1994) was inspired by the work of Moscovici, taking up the idea of the "figurative core" and enriching it. As Abric (2005) has observed:

> The figurative core will provide the framework for categorizing and interpreting new information reaching the subject [...]. The other elements of the representation will be retained, categorized and interpreted according to the nature of the figurative core.

Partly relying on the process of objectification, Abric was interested in the study of the internal organization of social representations from the structural approach which has since been tested by many researchers (Moliner 1989; Guimelli and Rouquette 1992). According to this approach, "the analysis of a representation, the understanding of its functioning, necessarily requires a double identification: that of its content and that of its structure", hypothesizing that "not only are the elements of social representation hierarchical, but every representation is organized around a central core, made up of one or a few elements which give to representation its meaning".

Abric insists on the importance of the structuring of the representation, making it possible to take into account a set of hierarchical elements and differentiating two zones: (1) the central core, a structuring and stable element, which determines the meaning and permanence of the representation and which is the bearer of consensus, and (2) the peripheral elements, which are organized around the central core and which are hierarchical, enabling both the maintenance of the representation and its adaptability to different contexts.

According to this theory, two representations will be different if and only if they are organized around two distinct central nuclei, to such an extent that identifying the content of the representation alone is not enough because it is also necessary to understand its organization. Peripheral elements, in contrast, have a more flexible character, as they are related to individual characteristics and the immediate environment. The position of the peripheral elements, more or less close to the central core, plays an important role in the meaning of representation. The identification of the central core and the peripheries, as well as the relations between the informational contents, makes it possible to put forward hypotheses on the processes of transformation of social representations apprehended and seen, according to Abric (2005), as "an (organized and structured) ensemble of information, beliefs, opinions and attitudes about a given object". The analysis of social representations can therefore make it possible to bring forward hypotheses on the transformation processes they undergo.

7.4. Methodology

To collect and identify the content of a social representation, we used the free and hierarchical evocations technique (Abric 2005) to assess the concept "how to produce differently", which was followed by a prototypical analysis (Vergès 1994) involving: (1) associative techniques based on the indexing of terms or groups of terms considered as indicators of the cognitive organization of individuals; (2) the free association phase, spontaneously indicating all the words or expressions that come to one's mind, enabling rapid access to the elements populating the semantic universe of the objects studied; and (3) the hierarchical structuring phase, aiming to classify the answers depending on the importance given to each term to define the object in question.

This methodology makes it possible to access spontaneous discourse on the object of representation and to establish the semantic universe of a given population. According to Abric (2003), it also enables "the updating of implicit or latent elements which could be drowned or masked in discursive productions". In particular, we have to identify the vocabulary associated with the notion studied, in this case "how to produce differently", and grasp the meanings that individuals express about it. Ultimately, this methodology provides an initial state of the representation's structure and its organization. To analyze the data collected, we used prototypical analysis and similarity analysis offered by Iramuteq software (R interface for multidimensional analyses of texts and questionnaires).

7.4.1. *Data collection*

Data collection was carried out during five sessions organized between September 16, 2014, and November 17, 2015, at the *École nationale supérieure de formation de l'enseignement agricole*. This collection took place in three phases corresponding to three distinct activities, and this was done during each of the sessions: (1) individual collection of the three words that best define the notion "how to produce differently", (2) collective work by groups of five to eight teachers from different disciplines who had to agree on six key words summarizing the notion "how to produce differently" and (3) a phase during which each teacher had to distribute five votes for the most representative notions on "how to produce differently" among the ones identified during the second phase.

7.4.2. *Populations*

Each session brought together between 53 and 87 teachers for a total population of 340 teachers, including 200 women (58.8% of the workforce) and 117 teachers (34.4%) from external agricultural education courses who were identified as "novices" (the 223 other teachers came from internal courses and were described as "experienced"). In terms of disciplinary affiliation, 34.0% of teachers qualified as "technical" because they came from technical and professional disciplines (food engineering, development of natural areas, landscaping, agricultural equipment, animal production, animal husbandry, aquaculture, hippology, horticulture, crop production, vines and wine).

The other teachers were grouped under the "non-technical" designation, among which 44.1% came from human and social science disciplines (economic and social sciences and management, socio-cultural education, history and geography, literature and history, modern literature, modern languages, documentation), 15.6% from scientific disciplines (biology and ecology, biochemistry, microbiology and biotechnology, mathematics, physics and chemistry) and 5.9% from other disciplines (physical education and sports, computer and multimedia technology).

7.5. Data categorization

Following the examination of the questionnaire, for each activity, we categorized the answers. For the first activity, relating to the individual representation of "how to produce differently", we collected 1,020 words or expressions (3 words per teacher × 340 teachers) and 1,700 words or expressions during the third activity (5 words per teacher for a total of 340 teachers).

The results were the subject of a previous publication in the *Environmental Education* journal (Asloum and Bedoussac 2020). We only recall the main lines here. Then, we present a categorization of the responses attributed to the techniques identified by the teachers as being the most characteristic of "how to produce differently" and indirectly referring to agroecology.

7.6. Results

Analysis of the tree of similarities for the entire population (Figure 7.1) shows an organization structured into four communities after completing Activity #1 ("environment" (31.1% of citations), "technical" (23.5%), "sustainable" (21.2%) and "change" (24.2%)) and into three communities after completing Activity #3 ("change" (45.6%), "education" (35.4%) and "environment" (18.9%)).

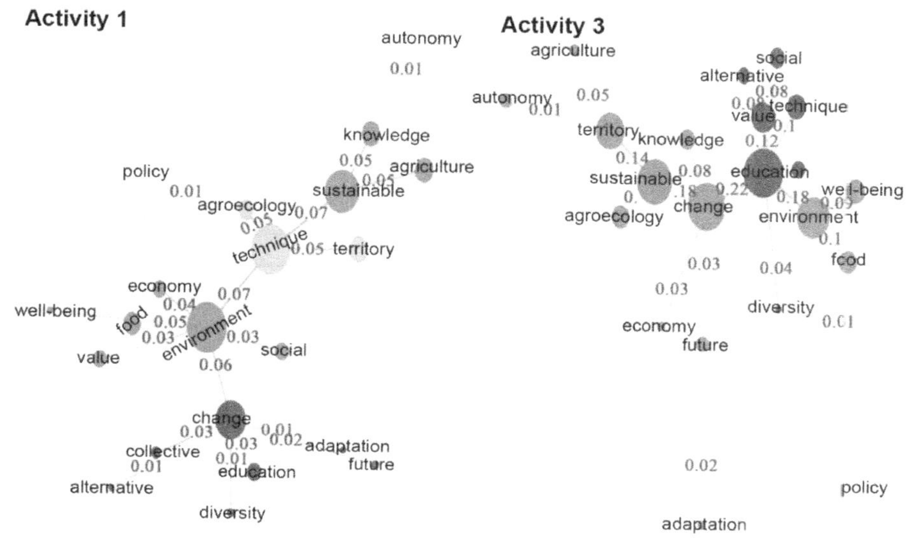

Figure 7.1. *Analysis of similarities according to the Fruchterman–Reingold algorithm (Fruchterman and Reingold 1991) regarding the individual representations of "how to produce differently" obtained during Activities #1 and #3 for the entire population (n = 340). For a color version of this figure, see www.iste.co.uk/barthes/biosphere1.zip*

COMMENT ON FIGURE 7.1.– *Disk size is proportional to the number of occurrences. A set of disks of the same color forms a community, identified by the "edge betweenness" Girvan–Newman method (2002). The blue numbers on the lines connecting two disks correspond to co-occurrence rates. Adapted from Asloum and Bedoussac (2020).*

For the whole population, the central core (Figure 7.2) resulted in six terms during Activity #1 ("technical", "environment", "change", "sustainable", "knowledge" and "agriculture") and four terms after Activity #3 ("change", "sustainable", "environment" and "technical"). It can be noted that the term order of the central core was almost completely reversed and that the terms "knowledge" and "agriculture" migrated, respectively, to the second periphery, as did the contrasting elements. The first periphery of the central core, considered by Abric (2005) as a potential zone for change, shows that "producing differently" was associated with the notions of territory and food during Activity #1, whereas at the end of Activity #3, we observe the appearance of the notions of education, value and well-being.

Activity 1

4.8% ≤ Frequency	Central core (Rank ≤ 2)	%	Rank	First periphery (2 < Rank)	%	Rank
	technique	12.6%	2.0	territory	5.7%	2.1
	environment	12.3%	2.0	food	5.4%	2.2
	change	10.5%	2.0			
	sustainable	9.3%	1.7			
	knowledge	5.7%	1.9			
	agriculture	5.6%	1.8			

Frequency < 4.8%	Contrasting elements	%	Rank	Second periphery	%	Rank
	agroecology	4.3%	1.7	economy	3.9%	2.4
	education	4.3%	1.9	social	3.6%	2.1
	value	4.1%	2.0	collective	3.0%	2.3
	alternative	1.2%	1.9	future	2.2%	2.2
				diversity	1.9%	2.1
				well-being	1.8%	2.3
				adaptation	1.2%	2.1
				policy	0.9%	2.6
				autonomy	0.6%	2.3

Activity 3

4.8% ≤ Frequency	Central core (Rank ≤ 3)	%	Rank	First periphery (3 < Rank)	%	Rank
	change	10.9%	2.7	education	12.3%	3.3
	sustainable	9.8%	2.5	territory	7.9%	3.1
	environment	9.1%	2.9	value	7.1%	3.2
	technique	5.2%	3.0	well-being	5.4%	3.4

Frequency < 4.8%	Contrasting elements	%	Rank	Second periphery	%	Rank
	agroecology	4.4%	2.2	food	4.3%	3.4
	alternative	3.0%	2.8	knowledge	3.9%	3.2
	agriculture	1.9%	2.1	social	3.7%	3.4
	diversity	1.1%	2.9	future	3.6%	3.2
	policy	0.1%	3.0	collective	3.1%	3.2
				economy	1.6%	3.3
				adaptation	1.1%	3.3
				autonomy	0.5%	3.2

Figure 7.2. *Prototypical analysis of the individual representations of "how to produce differently" obtained during Activities #1 and #3 for the entire population (n = 340) Adapted from Asloum and Bedoussac (2020)*

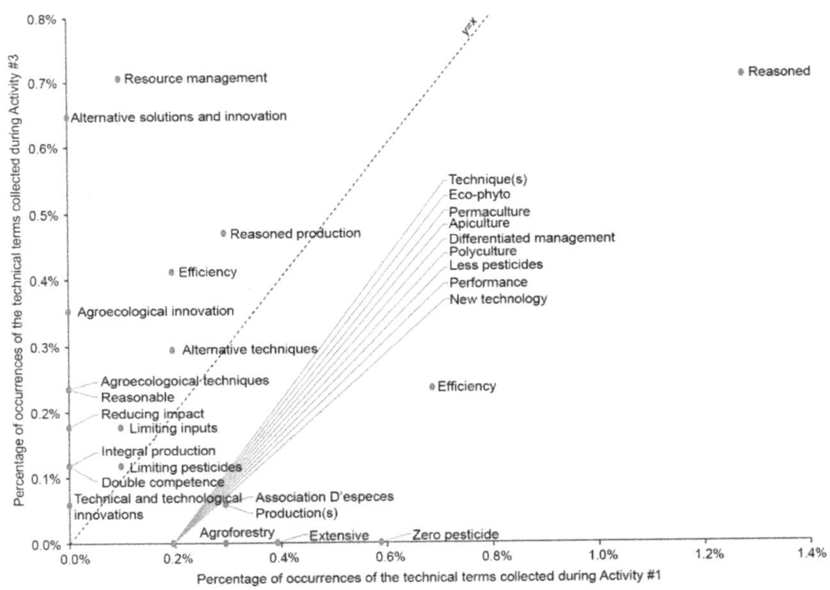

Figure 7.3. *Percentage of occurrences of the technical terms collected during Activity #3 as a function of those collected during Activity #1 for the entire population (n = 340)*

COMMENT ON FIGURE 7.3.– *Only 30 terms are represented, corresponding to the 19 most quoted terms during Activity #1 and the 18 most quoted terms during Activity #3. The percentages correspond to the number of occurrences of all the words or expressions collected (1,200 for Activity #1 and 1,700 for Activity #3).*

Some of the techniques mentioned by the teachers (Figure 7.3) were relatively accurate during Activity #1 (such as agroforestry, species associations, polyculture, beekeeping or even permaculture), whereas in Activity #3, they were replaced by more comprehensive notions, highlighting innovation and alternatives (see, for example, expressions such as "alternative solutions and innovation" or "technical and technological innovation"). In both cases, however, we observe the central place given to techniques aimed at reducing inputs and phytosanitary products. Finally, we can note the singular positioning of the term "reasoned", which was mentioned twice less during Activity #3 compared to Activity #1.

7.7. Discussion

The lexical analysis of the answers provided in Activity #1 shows that the representational field was largely dominated by the term "technical" in association with practices more respectful of the environment. A reference to the social and economic dimensions of sustainable development only appeared in the second periphery. During this same Activity #1, the answers proposed by the teachers emphasized the need to change agricultural practices in connection with the acquisition and transmission of knowledge through education.

The order reversal of the terms of the central core between Activities #1 and #3 testifies to a modification of the initial representations on "how to produce differently". We think that this reversal was the product of the exchanges between teachers made possible by Activity #2. Thus, the final representations of the teachers ended up being broader and more comprehensive, suggesting that teachers did not merely define the notion of "how to produce differently". In particular, the fact that the term "education" occupied a strong place in the first periphery suggests that teachers projected themselves into their teaching mission. Apart from their contribution to changing agricultural practices, teachers also aim to adopt behaviors that will enable "the participation and preparation of students for the present and the uncertain world to come" (Lange 2015).

The accuracy of the techniques quoted in Activity #1 suggests that teachers initially perceived agroecology as a juxtaposition of a set of practices, whereas at the end of Activity #3, teachers rather focused on the purpose than the techniques stricto sensu, all of which have to be combined in a coherent and systemic way. This first technical representation of agroecology can be seen as a necessary passage to access the construction of a broader meaning of "how to produce differently" and echoes the difficulty of translating agroecology into concrete techniques. In this sense, the technique would not be limited to the critical thinking of teachers, but on the contrary, it would be a means for unveiling it, joining in a certain way the epistemological considerations already established by Heidegger (1958) and Simondon (1989) regarding the technical question.

This change in the teachers' representations, from an initially technical thought towards a more global thought, was strongly related to the peer confrontation during Activity #2, which caused a broadening of representations, in particular due to the multidisciplinary composition of groups, without nonetheless resulting in a completely standardized vision of "how to produce differently". The heterogeneous representations of teachers vis-à-vis agroecology translated by the expression "to produce differently" are associated with the nature of the object and its complexity. Still, even though representations are elaborated within a social context, the social relationships in which the individual is led to join the group in order to appropriate the object of representation may change within the group, under the influence of interfering pressure (Moscovici 1984).

7.8. Conclusion

In agricultural education, program designers have to integrate issues arising in society and in the professional world even before knowledge and techniques have become stable. In addition, programs "no longer define the notions or concepts which are the subject of teaching so precisely, but rather constitute benchmarks" (Simonneaux and Simonneaux 2014).

While teachers have to cope with emerging concepts for which they have not necessarily been trained, our results show that their vision remains highly technical. Teachers rely on referenced techniques (agroforestry, intercropping, permaculture, polyculture breeding, etc.) aimed at reducing inputs – and, indirectly, the protection of the environment – while the other pillars of sustainability remain at the margin.

Agricultural practices are the result of a heritage constructed from standardized references essentially based on conventional agriculture, whose success, as Mayen (2016) has pointed out:

> tends to dismiss and neutralize much of the complexity of living processes, sets standards of practice which may not take into account local situations and contexts, cares little about the environment and natural resources.

Thus, unlike the techniques the teachers refer to, agroecology requires increasingly complex combinations of knowledge and practices. However, as Mayen (2016) has also pointed out:

> The challenge posed by agroecology is not just a knowledge challenge, in the sense of producing knowledge to meet the agriculturalists' needs for action and innovation. It is a challenge for learning and for educating the ability to think.

The interest of the questionnaire, even though based on declarative information, makes it possible to grasp what teachers understand about "producing differently" and the associated practices and techniques. In a transformation context driven, among other things, by the new framework plan "Teaching How to Produce Differently 2" (2019), the appropriation of concepts by teachers cannot do without training support.

These transformations also invite teachers to adopt a reflective posture vis-à-vis situations and techniques. However, the new generation of teachers we have been training for over 10 years now shows that they increasingly integrate the question of respect for the environment, certainly because of the growing importance given to the protection of the environment in society and their rising awareness during the initial training. Eventually, to enrich and develop the teachers' representations on "how to produce differently", it is necessary think in the long term, to ensure the anchoring and sustainable mobilization of agroecology in their professional practices.

7.9. References

Abric, J.C. (1994). L'organisation interne des représentations sociales : système central et système périphérique. *Structures et transformations des représentations sociales*, 73–84.

Abric, J.C. (2003). La recherche du noyau central et de la zone muette des représentations sociales. *Méthodes d'étude des représentations sociales*, 296.

Abric, J.C. (2005). *Méthodes d'étude des représentations sociales*. Éditions Érès, Toulouse.

Altieri, M.A. (1987). *Agroecology: The Scientific Basis of Alternative Agriculture*. Westview Press, Boulder.

Asloum, N. and Bedoussac, L. (2020). Analyse des représentations sociales des enseignants.es du "produire autrement" vis-à-vis des directives ministérielles. *Éducation relative à l'environnement : Regards – Recherches – Réflexions*, 15(2).

Béguin, P., Dedieu, B., Sabourin, E. (2011). *Le travail en agriculture : son organisation et ses valeurs face à l'innovation*. L'Harmattan, Paris.

Bellon, S. and Doré, T. (2019). *Les mondes de l'agroécologie*. Quae, Versailles.

Bellon, S. and Ollivier, G. (2012). Agroécologie en France : l'institutionnalisation d'utopies. In *L'agroécologie en Argentine et en France : regards croisés*, Goulet, F., Magda, D., Girard, N., Hernandez, V. (eds). L'Harmattan, Paris.

Flament, C. (1981). L'analyse de similitude : une technique pour les recherches sur les représentations sociales. *Cahiers de psychologie cognitive*, 1(4), 375–395.

Francis, C., Lieblein, G., Gliessman, S., Breland, T.A., Creamer, N., Salomonsson, L., Helenius, J., Rickerl, D., Salvador, R., Wiedenhoeft, M. et al. (2003). Agroecology: The ecology of food systems. *Journal of Sustainable Agriculture*, 22(3), 99–118.

Fruchterman, T. and Reingold, B. (1991). Graph drawing by force-directed placement. *Software: Practice and Experience*, 21, 1129–1164.

Garnier, C. and Sauvé, L. (1999). Apport de la théorie des représentations sociales à l'éducation relative à l'environnement – Conditions pour un design de recherche. *Éducation relative à l'environnement : Regards – Recherches – Réflexions*, 1, 65–77.

Gillet, G., Redonnet, J.M., Galaup, M. (2017). Autonomie du sujet et investissement de soi dans le cas de prescriptions discrétionnaires. In *Quatrième Colloque International de didactique professionnelle. Entre pressions institutionnelles et autonomie du sujet : quelles analyses de l'activité en situation de travail en didactique professionnelles*. Villeneuve d'Ascq, 6–8 June.

Gliessman, S.R. (1990). Agroecology: Researching the ecological basis for sustainable agriculture. In *Agroecology. Ecological Studies (Analysis and Synthesis)*, Gliessman, S.R. (ed.). Springer, Berlin.

Guimelli, C. and Rouquette, M.L. (1992). Contribution du modèle associatif des schèmes cognitifs de base à l'analyse structurale des représentations sociales. *Bulletin de Psychologie*, 45, 196–202.

Heidegger, M. (1958). *Essais et conférences*. Gallimard, Paris.

Jodelet, D. (1997). Représentation sociale : phénomènes, concepts et théories. In *Psychologie sociale*, Moscovici, S. (ed.). PUF, Paris.

Lange, J.M. (2015). Éducation et engagement : la participation de l'École à relever les défis environnementaux et de développement. *Éducation relative à l'environnement*, 12.

Mayen, P. (2016). Connaître et penser, le défi d'intelligence des pratiques agroécologiques. *Agronomie, environnement et sociétés*, 6(2), 167–175.

Moliner, P. (1989). Validation expérimentale de l'hypothèse du noyau central des représentations sociales. *Bulletin de Psychologie*, 41, 759–762.

Moscovici, S. (1984). Introduction : le domaine de la psychologie sociale. *Psychologie sociale*, 2, 60–78.

Simondon, G. (1989). *Du mode d'existence des objets techniques*. Aubier, Paris.

Simonneaux, L. and Simonneaux, J. (2014). The emergence of recent science education research and its affiliations in France. *Perspectives in Science*, 2(1/4), 55–64.

Stassart, P.M., Baret, P., Grégoire, J.C., Hance, T., Mormont, M., Reheul, D., Visser, M. (2012). L'agroécologie : trajectoire et potentiel pour une transition vers des systèmes alimentaires durables. In *Agroécologie. Entre pratiques et sciences sociales*, Van Dam, D., Nizet, J., Streith, M. (eds). Educagri, Dijon.

Tomich, T.P., Brodt, S., Ferris, H., Galt, R., Horwath, W.R., Kebreab, E., Leveau, J.H., Liptzin, D., Lubell, M., Merel, P. et al. (2011). Agroecology: A review from a global-change perspective. *Annual Review of Environment and Resources*, 36, 193–222.

Vergès, P. (1994). Approche du noyau central : propriétés quantitatives et structurales. *Structures et transformations des représentations sociales*, 277, 233–253.

Wezel, A., Bellon, S., Doré, T., Vallod, D., David, C. (2009). Agroecology as a science, movement or practice. *Agronomy for Sustainable Development*, 29, 503–515.

8

Collective Skills from Partnerships Between Protected Areas and Teachers

8.1. Introduction

A territorial dynamic is currently becoming manifest by an inflation of partnership projects in which the school frequently takes part. Following a deconcentration logic, a partnership was sealed between the French Natural Regional Parks (NRPs) and the Ministry of National Education. For each park, there was an agreement ratified by the NRP in question and the National Education office present in the territory. This link between educational establishments and territorial actors matches patrimony-based formal education with non-formal education. This encounter signals an evolution in the relationship with experiential knowledge (Charlot 1997; Khalali et al. 2019), which is precisely the approach favored by the NRPs. The acquisitions and knowledge collected in the field, outside the classroom, lead to a partial modification of the school form (Vincent 1994; Maulini and Perrenoud 2005). They are likely to modify the relationship with knowledge (Khalali et al. 2019) because the "lived" experience is added to simple factual knowledge. The non-formal educational approach is based on the interaction and direct contact between the learner and the territory, enabling the learner to acquire contextualized know-how and knowledge. Formal education is defined as a "metaphorical meeting place between the student and the source of knowledge, between the receiver and the sender of the training contents, in a new relationship, built to express the meaning of knowledge" (Garnier 2014).

Here, we shall not develop the epistemological questions raised by the school form. Our study focuses on partnership practices among teachers and guest speakers from an NRP who have experience in patrimony education projects. In order to

Chapter written by Sylviane BLANC-MAXIMIN.

properly grasp the present issues and their impact on educational activities, it is essential to describe the professional context of each partner and the values they are supposed to embody and/or transmit.

The current goals pursued by all the French Regional Natural Parks can be broken down as follows: innovating in favor of sustainable development, valorizing patrimony, supporting the local economy and educating about the environment (source: official French Regional Natural Parks website). An NRP operates with an economic, social and educational synergy in a protected area. Patrimony occupies a prominent place, and it is essential to valorize it. This valorization of territories and protected areas is organized around at least two axes: a tourist economy and an educational dynamic.

This approach is supported by values specified in charters. For example, the one from the Provence–Alpes–Côte d'Azur (PACA) region is based on a principle of pooling knowledge and skills in order to enrich practices and nurture the quality of projects, innovation and creativity, in accordance with the PACA NRP territorial education charter (ETT file). "The park works with the territory's actors, who recognize themselves in the educational principles it defends. It ensures that the projects carried out are long-term, and favor active educational approaches."

In concrete terms, an NRP has an address book of self-employed educational actors, who – depending on the region – have different skills (e.g. mid-mountain guide, botanist, fishing guide, potter, musician, etc.), and of partner associations (patrimony protection, hiking, etc.), which are contacted on the basis of their proposals for educational projects involving primary and secondary school teachers. Guest speakers are neither civil servants nor recruited on a permanent basis (with an employment contract). Their status is similar to that of intermittent workers (paid "by the task"). An NRP invites them depending on the needs expressed by the teachers who wish to commit their classes to partnership projects.

On the part of the Ministry of National Education, institutions invite teachers to develop educational pre-projects in anticipation of a meeting between the future partners. Within this framework, and depending on the contents, the NRP invites guest speakers identified as experts in a field. Teachers possess professional skills listed in a business repository (MENESR 2013b), which has a special mention of "cooperating with schooling partners". Another institutional text – the French Circular No. 92-196 from July 3, 1992, on participation of external speakers in teaching activities in nursery and elementary schools – explicitly describes the reasons that prefigure the presence of guest speakers in the formal school setting. This is legitimized by expertise in a specific area that the professor lacks. The presence of a guest speaker remedies this but nonetheless raises questions as to the professional identity of teachers.

To fully grasp the stakes of this partnership, we are going to explore this concept further and identify the salient elements for its implementation. We approach this phenomenon from the side of joint work between the guest speaker/teacher, which engages specific professional positions. Instead of thinking about it in terms of shortcomings (particularly on the teacher's side), we try to find out whether these professional encounters can be an opportunity to enrich professionals in a specific way, by engaging skills of a collective nature.

8.2. The educational partnership

In 2006, Mérini brought to light the expectations a partnership situation raises: openness to the other, the dimension of the real and the authenticity of practical implementation in a learning environment (the school) more traditionally oriented towards "formalized" knowledge, strongly influenced by didactics. In the case of patrimony-based education with an axis on the territory, teachers position themselves in favor of "a school 'integrated' into a territory, while remaining open to territorial partners" (Lange 2018). This transcends the mere orientation of a school open to the territory, guided by a teacher, as was the case, for example, with the class-outings (*classe-promenade*) initiated by Célestin Freinet.

Not only does the intervention of an actor outside the school positioned as an expert who must fill in the teacher's gaps bring enrichment to all (students and teacher-educator), it also raises tensions. This working modality is part of the dual-education culture established between school practices and reference social practices that exist, for example, in patrimony-based education. The richness and difficulty of partnership work stem from the interaction between the values specific to each partner, involving an alteration (Ardoino 1993) of the two cultures that leads (Clanet 1990) to the development of the concept of interculturality. The partnership situation is potentially based on a rich interaction between the actors and on the updating of new professional values, with cross-views on complex themes with different methods. However, this interaction is delicate because it can generate inter- and intrapsychic conflicts of professional identities which call for negotiations and renormalizations among the actors (Schwartz and Durrive 2003). In this intercultural encounter between professionals, the 1994 INRP study conducted by Zay identified the presence of strong issues for teachers: disciplinary, ideological, and professional identity issues (the teacher can no longer be defined by the knowledge of contents, but rather expects that the collaboration with the other will yield more than what is already known) and modification of teaching practices, evaluation methods and self-training or mutual training (Dupont 2010; Eneau 2016). Training can take the form of professional skills. A negotiation skill can involve the acquisition modality between professionals. Negotiation primarily concerns the project's programmatic aspect (Boutinet 2005): financial arrangements, schedule and duration of the actions, and the choice of the finished product as specified by a set of goals to be achieved during identified tasks or situations. Thomazet and Mérini (2014) state that it is undoubtedly at this point that the issue of partnership comes into play, as we have been able to define it; that is, a negotiated action taking into account the interests of each, with a view to greater efficiency in the student's learning career (Mérini 2007). This action presupposes skills in negotiation, contract formalization and decision-making regulation. However, this architecture work often leaves in the dark the discussion on the dimension of the educational process present in a project (Boutinet 2005). In the three major axes that seem to underpin collaborative practices (Thomazet and Mérini 2014), the two axes of elaboration, action planning and communication, often obscure the third axis (evaluation), which is fundamental in the learning and professional development processes of educators and teachers. In point of fact, careers are not monolithic and do not strictly result from prescriptions, be they regulatory (primary prescription) or related to training (secondary prescription). Professionals invent and reinvent their professions: "At each moment, a human being is full of unrealized possibilities" (Schneuwly and Bronckart 1985). This is because the activity performed and the real activity do not overlap (Clot 2007). The identification of the different configurations of collaborative practices (Thomazet and Mérini 2014) leads to a typology: partnership practices, collaborative exchanges operationalizing decisions (autocratic, subcontracting, others), collaborative mediation exchanges and joint practices.

8.2.1. *Partnerships in national education*

The notion of a partnership appeared in France in the 1990 framework law, which introduced the concept of skill in the field of national education. The skills logic (Zarifian 1999), transposed from the business world to the school environment since 1991, encourages teachers to apply professional skills in the exercise of their profession. In this regard, a professional skills repository governs the professionalization of teachers. The teacher has to "cooperate with the school partners" and in particular with "cultural structures and socio-economic actors, by identifying the role and action of each of these partners" (MENESR 2013a).

The school partnership aims to introduce the presence of expert partners in a specific field to the school's usual learning system. At the practical stage, it appeals to different postures collective in nature: coordination, collaboration and cooperation (Marcel et al. 2007) with the teacher and with the guest speaker.

Partnership in an "education for" project with an axis on the territory calls for several of these configurations. Educational actors are required to develop skills in negotiation, contract formalization, regulation for joint decision-making and concrete implementation. Hence, the concepts of collective intelligence (Lévy 1994) and collective skill (Wittorski 1997; Dubois and Retour 1999; Astier 2006), initially built on the interdependence of specialists and of experts, should be able to benefit the "education for" valorizing territorial projects (as in patrimony-based education). In fact, in the current framework of teaching programs, these are positioned as rational islands of interdisciplinarity (Fourez 1998). In the territory's context, the "education for" calls for the professional skills from NRP guest speakers (members of associations, protected areas, etc.). To some extent, this jostles the teaching professions and, in broader terms, calls into question the role of the school and the learning that takes place therein. Not only does partnership work involve meetings of knowledge between actors, it is also affected by conflicts of values: those transmitted by the actors themselves and those derived from patrimony-based education, which conveys contradictory values (Barthes et al. 2015). While these tensions can generate conflicts, they can also invite actors to develop skills of a collective nature.

8.2.2. *A partnership for creating collective professional skills?*

Research work (Blanc and Eymard 2010) has made it possible to transpose the notion of collective skill from the business world (Le Boterf 2000) to the sphere of primary education. In the world of education, there are two skills which are collective in nature: a collective skill (*compétence collective*) and the collective's skill (*compétence du collectif*), each characterized by indicators of their presence.

<table>
<tr><th colspan="2">Indicators signaling the presence of a collective skill</th></tr>
<tr><td rowspan="2">Ensuring mutual training</td><td>Sharing in-depth expertise in a field with partners</td></tr>
<tr><td>Building shared representations of pedagogies, learning, evaluation</td></tr>
<tr><td rowspan="2">Overcoming compartmentalization</td><td>Exchanging ideas, "recipes" and artifacts for patrimony-based education</td></tr>
<tr><td>Sharing didactic and pedagogical documents</td></tr>
<tr><td>Collectively achieving the expected results</td><td>Carrying out the actions planned in the project
Evaluating together the project's effects and reflecting on the means for evaluating the students' achievements</td></tr>
<tr><td>Establishing a fair workload</td><td>Distributing the workload required to bring a project to fruition</td></tr>
<tr><td>Acting by making and implementing decisions which have been previously determined in concert</td><td>Setting common goals when drafting the project
Co-writing the project</td></tr>
</table>

Table 8.1. *Transposition of indicators of collective skill for a partnership between NRP guest speakers and teachers*

Finally, a synthesis from these indicators (Blanc and Eymard 2010) and the work of Marcel et al. (2007) on action verbs (to coordinate, to collaborate and to cooperate) will be used for identifying the nature of NRP–National Education-type partnerships in the cases studied.

The collective skill, close to collective intelligence (Lévy 1994), favors the effective resolution of a problem with a view to reactivity and profitability for the organization, the collective and the individual. Projecting an "intelligence distributed everywhere, constantly valorized, coordinated in real time and leading to an effective engagement of individual skills" (Lévy 1994, p. 29), the collective intelligence summons and creates synergies between the skills of individuals, under the principle that no one knows everything, that everyone knows something and that all knowledge is contained in humanity. Cross-checks can be established with one of the goals of territorial intelligence (Girardot et al. 2014), aiming to offer partner actors a cognitive process and the tools for the co-construction of a collective intelligence for, together with, and carried out by the territorial community. This process materializes in the development of tools accessible to actors for developing, managing and evaluating sustainable development projects. It is characterized by professional learning of the hologrammatic type (everyone is a carrier of the whole).

The collective's skill is deployed in the capacity of the collectives to engage their resources without necessarily leading to the construction of a skill for each member. It is the quality of the action carried out (a project, for example) that matters. However, a certain porosity exists between these two types of skills: professional learning, which is not an indicator of the collective's skill, can take place later among the actors. Actors are not forced to develop a common operative image under the collective's skill if the treatment does not correspond to the values which affect them. Finally, this type of collective skill (linking the two types) complements an organizational perspective and an approach that valorizes human inventiveness, in harmony with their professional values and conceptions.

As a result, the multiplication of educational projects associating territory-based actors with teachers raises epistemological questions. The meeting between two educational worlds calls into question purposes, values and the school form. It calls into question the professional profiles of the actors and their professional development.

Within the framework of this text, we study what is at stake in situations involving partnership practices among territory-based "educational" actors from the perspective of patrimony education.

Collective skill: combination of the individual skills from the collective's members	**The collective's skill: juxtaposition and coordination of the individual skills from the members of the collective**
Cooperation	Coordination and collaboration
Developing skills other than individual ones: what we can do together but we cannot achieve on our own	
Combination, arrangement in the mathematical sense Bringing components together by arranging them in a certain order, with a view to a certain result	← Collection in the mathematical sense Synonym of an unarranged set
Synergy	← Coordination
Complementarity of specific skills asserted in a field that are engaged in a combinatorial way to solve a problem, to build a project or to carry it out	No asserted skills, but the association of the individual skills from each member makes it possible to carry out a project and to solve a problem
Learning attainment by the members of the collective. Some teach others and others teach the former: collective learning (mutual training) Professional learning in a hologrammatic dimension (each member has everything that has been decided and learned inside themselves)	There are probably some already learned elements. It is in the attainment of the completed product or project that experiential learning takes place
Stored in memory Cognitive benefit for project members Creator of professional and reference knowledge	Benefit thanks to the action conducted (but not necessarily putting it into perspective), integration by the members
Elaboration of shared representations (Astier 2006) Common operative picture Common operative language Accepting the existence of conflicts by managing them in a timely manner	Operative image not necessarily common Common operative language Avoiding conflicts or letting them linger in a latent state

Table 8.2. *Comparison of the characteristics of the two skills*

More specifically, we seek to know to what extent and under what conditions partnership practices between NRP guest speakers and teachers during patrimony education projects develop skills of a collective nature among the actors.

8.3. Three case studies in a labeled rural territory

This study is set in an isolated rural territory, identified by the Ministry of National Education for its socio-economic and cultural difficulties. It has been labeled as part of the priority education network in rural areas. Cultural patrimony entries are privileged here. The educational actors (teachers and guest speakers invited by the NRP from the Provençal baronies) have the project of involving the students in the discovery of patrimonial or potentially patrimonial, material and immaterial elements of their living territory, whose wealth is unsuspected by the students and their families. The educational policy of the NRP at the national level aims to develop a territorial anchoring among the pupils. Teachers aim to impart knowledge and skills recorded in the official programs in force issued by the Ministry of National Education.

8.3.1. *Descriptions of the projects undertaken in the three case studies*

8.3.1.1. *Case A: Patrimony-based education by developing the link to nearby patrimony sites*

Case study A took place in three multi-course elementary classes (from CP to CM2) located in three rural villages of less than 500 inhabitants belonging to a priority education network. The partnership project includes the intervention of a guest photographer invited by the NRP. The photographer wishes to link the three rural schools under one project, which could help remove the feeling of geographical isolation and above all could contribute to valorizing their favorite patrimonial places in this environment. Each student would have their own portrait photographed by one of their companions at a place in the village that they consider a common good to be preserved for future generations. For the youngest students (5–7 years old), the goal would be to have their photograph taken at a place in the village that is important to them, to be discovered by other children of the same age. For example, the students chose the cliffs (a climbing spot renowned at the European level), the bistro, an old lodge dating from 1796, the river and its Roman bridge, a meadow with sheep, a lavender field and a Saracen tower, while very young students chose the cabin and the fountain in the public square or the games in the town's public garden.

8.3.1.2. *Case B: Education in local medieval patrimony*

Three fifth year classes from a priority education rural college and an association for the valorization of patrimony were involved. The Middle Ages was the interdisciplinary theme chosen and several college teachers (French, ancient languages, plastic arts, physics, technology, history) took part in the project.

The regional natural park was appointed as a partner in the project under the French Association Law No. 1901. Its members were people passionate about the Middle Ages and aimed to enhance and valorize the medieval patrimony by means of a camp, which would help the students to discover ancient trades and activities through workshops on armament, fencing, archery and games of skill.

8.3.1.3. *Case C: Education in natural patrimony (the bees)*

This project partnered up two kindergarten medium- and large-section classes (students aged 4–5 years) from a rural school and a guest speaker regularly invited by the NRP. The speaker proposed "multidisciplinary activities where playfulness, the scientific, the sensory, the manual and the visual are involved" and wished to contribute "knowledge and understanding of the concepts related to the protection of nature and the preservation of the environment" (excerpts from the website homepage of the participant, who is self-employed). The theme for this project was bees' patrimony as a form of heritage and biodiversity that should be protected.

8.3.2. *Data collection methods*

To tackle our case studies and provide answers to our research question, we opted for several methodological tools. A study of the projects submitted to the NRP's educational service and to the National Education district made it possible to identify the level of collaboration between partners in project planning.

The observation of the sessions dedicated to the project and carried out in partnership aimed to collect information on the type of partnership implemented in front of the students.

Observation and note-taking during the interviews between the guest speaker and the teacher (before or after the session) focused on the quality of communication, negotiation and possible contract formalization.

The observations were collected by note-taking or by video recording of the interviews (if accepted by the two actors).

They concerned the nature of the content of the exchanges (practical technical coordination of activities, co-intervention organization during the sessions, anticipated sharing of each person's roles, document sharing, etc.), who would conduct the interview (importance of one actor over the other, equal role distribution), possible conflicts and negotiations, clarifications between the two educational actors and ways for measuring the project's progress.

They enabled the analysis of the pedagogical practices of the two educators during the students' oral and photographic productions and the evaluation methods for the students' achievements that one or both partners implemented. These elements were to inform us on the quality of the partnership and the potential professional skills that were engaged.

8.3.3. *Results*

8.3.3.1. *The study of projects*

8.3.3.1.1. Case A

The programmatic aspect of the guest photographer's project had a strong impact on the writing of the final project. The teachers' intentions and goals converged with the speaker's proposals. The skills the students would be able to work on were vaguely specified and very little adapted to the contents of the official teaching programs.

8.3.3.1.2. Case B

The "banners and coats of arms" project was presented as an interdisciplinary project in relation to language proficiency. The project's title presented by the teachers suggested a fairly accurate split between the school activities and external interventions.

The project's main part was quite voluminous and written by the teachers, who solely focused on the learning contents (integral study of a book (*Yvain or the Lion Knight*); study of courtly love poetry; design of a coat of arms; drawbridge operation, the model of the fortified castle; society in the Middle Ages), addressing the Middle Ages theme from the perspective of each school discipline. The perspective was multidisciplinary, rather than interdisciplinary.

In the project, the place reserved for the goals and actions contemplated by the partners was succinct and presented as a lively staging of life in the Middle Ages at a camp. A medieval festival (see college details) was planned to round off the project. Another component envisaged an end-of-year trip on the theme of the Middle Ages in Paris.

A clear partition existed between the school, the "playful" aspect and the "live" experience proposed by the association for the promotion of medieval patrimony. During the lessons, the school form (exercises, essays) was thoroughly preserved and juxtaposed with the activities offered in the emblematic setting of the town castle. The project came to a close with a large medieval festival partly reserved for students and partly open to the general public, including animation for the local territory. The municipality provided substantial material assistance.

8.3.3.1.3. Case C

The reading of the project *Abeilles en conte, abeilles en danger* ("Bees in Tales, Bees in Danger") showed real co-writing between the teachers' expectations (whose goals and the skills they aimed to teach were highly detailed and compliant with the official program expectations) and what the speaker had to offer. The part reserved for the speaker was consistent with the one reserved for the teacher. The skills and situations proposed were complementary and identical to those described by the teachers. A description of the sessions showed the learning progression, the concepts worked on, the co-interventions during certain sessions and the involvement of the students' families in the project.

8.3.3.2. *Class observations*

Three forms of partnership were identified. Some developed practices of coordination, collaboration or cooperation and indicators of the collective's skill or of a collective skill were identified. Potentially, they could lead to the professional development of the actors.

8.3.3.2.1. Case A: Different partnership practices

School one: Confusion of professional roles

The teacher let the guest speaker do the course on their own, based on contents which nonetheless fell under the teacher's responsibility: at times, the photographer played the role of the teacher, without the teacher worrying about this. During debriefing or preparation for the following session, the speaker took the initiative for the exchanges and the teacher seemed to agree with the statements. There was no conflict, but rather a soft consensus set in. From that moment onwards, the speaker started giving out directions and as the project unfolded truly became the project's leader. This total delegation of the project's progress to the speaker did not open up any prospects for the engagement of collective skills.

School two: "Positive" conflicts

The teacher wished to keep control over what was offered to the students within the school framework to keep the initiative and so that these requirements were met by the guest speaker. The teacher wanted to retain decision-making power over what would be done and how, particularly in terms of evaluation of the students' achievements. The teacher did not intend to be told what to do and instructed the guest speaker to scrupulously respect the project's specifications. This resulted in situations of conflicting exchanges with the guest speaker, which led to progress on both sides. The photographer remained circumspect for a time, but then realized that he had something to gain because the partner's requirements included areas for improving his own interventions.

School three: An absence of partnership

The teacher adopted a withdrawal attitude and was reluctant toward the project, which had practically been imposed on this school. She refused any interviews outside of class time. There was no regulation between the partners and the guest speaker alone was granted the project's management and implementation. The partnership turned out to be non-existent, as were the collective skills.

8.3.3.2.2. Case B: A "pre-square" logic

A partitioned-type partnership was set in a "pre-square" logic. Although coordination was granted from the point of view of organization, nothing was addressed in terms of cognitive learning. The actions conducted by the association mostly seemed to focus on promoting the project among the elected officials, the students' families and the territory's inhabitants. The speeches of the two partners converged in the organization of the medieval festival. However, a real partition between school knowledge and practical knowledge on medieval combat techniques and on local medieval gastronomy was established. The partners' exchanges mainly addressed the festival's organization (security conditions, municipal authorizations, invitations, etc.).

8.3.3.2.3. Case C: Partnership as a vector of collective skills

Collective skill indicators were identified. The partners shared reference knowledge and teaching practices and learned from each other's experience. They had a common operative image and communicated regularly using a shared language. We noted the presence of time for reflexive analysis and systematic regulation before and after each project session. The initial project was systematically refined in accordance with student progress, and as the project evolved, the partners adopted a shared language. There were numerous pedagogical and didactic exchanges, based on the sharing of documents, the exchange of teaching material and the search for creativity in the situations proposed to the students. The teachers shared with the guest speaker the work done in class from the previous session, illustrating the progress achieved. For example, the guest speaker used a rhyme on bees during the session's transition time.

Each teacher agreed on the co-intervention modes with the guest speaker. For example, a teacher was able to step back for 20 minutes in order to observe the attitudes and activity of two hyperactive students and one usually inhibited student. Before the session, the guest speaker asked the teacher and the ATSEM for permission to perform certain tasks during the session.

8.4. Presence of a collective skill and of the collective's skill

8.4.1. *Case A in school one*

Case A in school one presented some rare indicators of the presence of the collective's skill. There was a benefit as a result of the action taken, but it was not necessarily put into perspective, subjected to reflection or integrated by the actors. This resulted in an operative image which was not necessarily shared by all. The duration of classes and interviews revealed that one of the teacher's goals was to avoid any kind of conflict. Despite the fact that the teacher did not refuse any of the photographer's proposals, she did not make any proposals of her own nor did she show any particular initiative. On the contrary, her sole participation was limited to the "smooth" achievement of the project by scrupulously respecting the guest speaker's directives. Joint reflection on the actions carried out never took place. As a result, the common operative image was quite vague in the teacher's words to the students.

8.4.2. *Case A in school two*

Case A in school two showed five indicators of the presence of the collective's skill and five indicators of the availability of a collective skill. It revealed the development of skills other than individual ones. The actors successfully identified what could be done together that could not be achieved on their own. This process took place by combination, arrangement in the mathematical sense (bringing ideas together, arranging the actors' proposals in a certain order, with a view to a certain result). Skills were brought into synergy through a common operative language. Although conflicts existed, they were managed in a timely manner. There were probably learned elements in both types of actors. This partnership led to a skills synergy between the two parties involved and encouraged the use of a common operative language. It was in the attainment of the completed product or project that experiential learning took place.

The role of conflicts was a remarkable element because their management in a timely manner led to improvement in the teaching situations and practices for students. Thanks to the conflicting interactions between partners – each wanting to claim or retain prerogatives – synergizing the skills from each (e.g. in terms of how to evaluate the students' achievements) produced great benefit in the actions carried out. The partners pooled their knowledge, know-how and respective attitudes within the framework of a criterion-referenced evaluation process co-constructed with the students. This approach was the result of the combination of partner skills. The asserted skill of a collective nature was nonetheless reduced to this example. There was no final evaluation of the project in terms of the quality of the effectiveness and the interest of a partnership. The common operative image was built step by step throughout the project's course, but it was the combination of the individual skills from each person that made it possible to carry out the project and to solve a problem.

8.4.3. *Case A in school three*

Case A in school three represented a case of virtual absence of partnership. The only identifiable indicator was the will not to create any conflicts, reflected by a complete laissez-faire attitude towards the guest speaker during his interventions in class, which were neither prepared jointly nor regulated with the teacher. For the partners, the question of evaluating this procedure was never raised.

8.4.4. *Case B*

Case B offers a large number of indicators of the presence of the collective's skill. Coordination and collaboration were limited to scheduling the project, in particular, the organization of the medieval festival which, on the one hand, rounded off the project and showed the complementarity of the educational actors involved in the project and, on the other hand, functioned as a showcase for the intervening association. It is clear that both parties had to develop skills other than individual ones: what could be done together that could not be achieved on their own. For example, the organizational skills for an event of such magnitude were particularly engaged: a teacher appointed by her colleagues was in charge of scheduling the program and submitted it to the head of the association. As we can see, the idea was to gather a set of skills which could then be coordinated. It was the combination of the individual skills from each that made it possible to carry out a project or to solve a problem. The intended benefit was solely oriented towards the success of the action carried out. The exchanges that we witnessed in no way related to the content of the learning offered by each partner, nor to the teaching methods proposed. This engagement of diverse skills was neither put into perspective nor integrated by the members, who did not wish to share their knowledge nor their skills. The guest speakers were identified by teachers as service providers. The remarks collected during the exchanges, which preceded the interventions in class or during the medieval festival, only showed a complementarity "to enable students to discover the literature of the Middle Ages and to playfully put themselves in the shoes of a knight", as stated by the ancient languages teacher. The French teacher supplemented the words of her colleague: "We are working on a knight's story, so it would be really good if you could make them use weapons, observe or take part in combat. Everyone will be able to understand, it will make sense."

8.4.5. *Case C*

Case C presented all the indicators of the presence of collective skill. Cooperation existed throughout all the project's stages and included knowledge sharing (types of foraging insects, pollination of flowers by bees, life in a hive and the roles of the different bees, the dangers awaiting them), an exchange of teaching methods (the guest speaker designed artifacts – games and a short animated film – suitable for young students and the teacher compiled tales about bees) and the exchange of practices (recognizing the parts of a bee's body, recreating it by means of land art for the guest speaker, using a stuffed animal, cut-outs and collages of the different parts of the bee's body and finally having a bee drawn for the teacher). Professional learning was achieved by the members of the collective (the two teachers and the guest speaker). There was mutual informal training, with some partners teaching the others and vice versa. A hologrammatic dimension (everyone has inside of them everything that has been decided and learned) was present. During an interview between the partners before a session, the teacher explained that she had tried to reproduce part of the guest speaker's session with the students who had been absent and that she had succeeded. The guest speaker replied: "It's funny, I integrated an extract from the tale that you made me discover in an elementary course A2, and I was able to create an entry to reflect on the loss of biodiversity". A co-intervention was negotiated and scheduled for each session when the speaker was present (organization in a certain order, so as to obtain a certain result). For example, the guest speaker and each teacher brainstormed ways to assess the progress of kindergarten students. Everyone tried to identify evaluation scenarios and the teacher had the mission (which also corresponded to her prerogatives) to collect traces of learning. It was not only knowledge or skills that were identified. A joint decision was made regarding the role of each. During a session led by the guest speaker, a teacher focused exclusively on the behavior of four students who, in the class, were not yet able to adopt attitudes involving the development of skills other than individual ones: what we can do together that we cannot achieve on our own. The two partners discussed the results of this observation and how to spark the interest of these students and how to harness it.

8.4.6. *Synthesis*

To summarize, the indicators of the presence of a collective skill concerning case C were numerous. In case A in school two, the indicators identified clearly denoted the presence of the collective's skill and some of them fell within the framework of a collective skill. Traces of the collective's skill existed in case B. As for case A in schools one and three, nothing proved the existence of this type of skill.

These case studies from the same institutional partnership showed different levels of partnership which may or may not offer opportunities for professional development. The obstacles identified primarily concerned the goals of educational actors. Some of them only aimed for the success of the action to the detriment of possible sources for professional learning which the activity's design and implementation could have enhanced. Others consumed the external interventions in a utilitarian manner and only got poorly involved (if at all) to acquire a little more expertise. Finally, some teachers seized an opportunity with the guest speaker in the situations proposed by this type of project (construction, production, intermediate reflexive times, etc.).

8.5. Conclusion

To answer our question on the favorable conditions for the development of collective professional skills for the actors involved in partnership practices between NRP guest speakers and teachers within the framework of patrimony-based education projects, we studied a few cases in rural areas in a protected area. The following points came as a result.

For a partnership between formal and non-formal education and learning to be fruitful, the partition of actors should, at least, be clearly identified: bringing to light and assuming the educational and ethical goals of each partner, acting in function of the skills contributed by each partner and taking into account the institutional goals prescribed for school teachers. Training in professional and associative partnerships involves acquiring and engaging cooperation and collaboration skills between the actors of a school educational project. Patrimony-based education necessarily spans long time periods that only a partnership between formal and informal structures can assume.

Partnership practices and their implementation in patrimony-based education projects are numerous and the variety of their efficiency can impact the learning of students and educational actors. The idea is not to standardize the practices of partners, but rather to enable them to learn professionally how to work together. The French school conception of partnership understands it in terms of assistance (the guest speaker is considered as an expert who comes in aid of the teacher) and subordination (the guest speaker merely represents an occasional help for the teacher), which is a major obstacle in itself. The relationship between partners is hierarchical: the school's own missions take precedence over the missions it cannot manage alone. Even though the Thélot report stipulates that only the school educates and that the other educational actors develop socialization, the cases we have studied contradict this assertion. If, in institutional terms, partners are considered as experts but with an assistantship status in a subordinate relationship towards the teacher, this means that, too often, educational projects are developed with partners, but not between partners. This poses the question of the evolution of this situation for student success and for the acquisition of new professional skills among the educational actors. We advocate for common training times for territory-based educational actors, in which case the territory would act both as the motive and the promoter. This track would certainly constitute an innovative training framework for National Education and the local partner, given its contextualizing dimension. In virtue of their characteristics, could the territories labeled as "protected areas" serve as the instigators for a new type of training validated by National Education? The bill for the reform of the school of the republic (2013) intended to orient students towards artistic and cultural paths, the future, health and citizenship, options that were not called into question by the 2019 law. The guidelines of these last two laws led to a multiplicity of "education for" projects, a sign of a school entrusted by society with the management (or delegation) of some of its contemporary problems. In addition, we are witnessing a pendulum movement between a school knowingly located outside of life, exclusively focusing on school learning to protect children from danger (Vincent 1994; Levrard and Delalande 2015), and a "contextualizing" school that acknowledges the local, environmental context, open to local realities and considering the territory as a support worthy of learning. This education contributes to conceiving the school as a singular experience of the world, lived in a collective and shared experience. Under these circumstances, instead of continuing to act under conditions of subordination and assistance, the collective of territory-based educational actors should be able to orient themselves towards collective intelligence (Lévy 1994) and territorial intelligence (Girardot et al. 2014) perspectives.

8.6. Appendix

Collective skill: combination of the individual skills from the collective's members	**The collective's skill: juxtaposition, coordination of the individual skills from the members of the collective**
Cooperation Case C	Coordination and collaboration Cases A2 and B
Developing skills other than individual ones: what we can do together but we cannot achieve on our own. Cases A2, B and C	
Combination, arrangement in the mathematical sense. Bringing components together by arranging them in a certain order, with a view to a certain result ← Cases A2 and C	Collection in the mathematical sense Synonym of an unarranged set Case B
Synergy ← Cases A2 and C	Coordination Case B

Collective skill: combination of the individual skills from the collective's members	The collective's skill: juxtaposition, coordination of the individual skill from the members of the collective
Complementarity of specific skills asserted in a field that are engaged in a combinatorial way to solve a problem or to build a project or to carry it out Case C	No asserted skills, but the association of the individual skills from each member makes it possible to carry out a project and to solve a problem Cases A2 and B
Learning attainment by the members of the collective. Some teach others and others teach the former: collective learning (mutual training) Professional learning in a hologrammatic dimension (each member has everything that has been decided and learned inside themselves) Case C	There are probably some already learned elements. It is in the attainment of the completed product or project that experiential learning takes place Case A2
Stored in memory Cognitive benefit for project members Creator of professional and reference knowledge Case C	Benefit thanks to the action conducted (but not necessarily putting it into perspective), integration by the members Cases A1, A2, B
Elaboration of shared representations (Astier 2006) Case C Common operative picture Case C Common operative language Cases A2 and C Accepting the existence of conflicts by managing them in a timely manner Case A2	Operative image not necessarily common Cases A1, A2, B Common operative language Case A2 Avoiding conflicts or letting them linger in a latent state Cases A1, A3

Table 8.3. *Collective skills and the collective's skill*

8.7. References

Ardoino, J. (1993). L'approche multiréférentielle en formation et en sciences de l'éducation. *Formation Permanente*, 25–26.

Astier, P. (2006). Apprendre et agir "ensemble". *Colloque SIFA-CREAD*, Rennes.

Barthes, A., Alpe, Y., Blanc-Maximin, S. (2015). L'éducation au patrimoine, un outil pour le développement local ? Positionnement critique vers l'instrumentalisation de l'éducation au service de la labellisation des territoires ? *Revue francophone du développement durable*, 3, 8–22.

Berryman, T. (2003). L'éco-ontogenèse : les relations à l'environnement dans le développement humain – D'autres rapports au monde pour d'autres développements. *Éducation relative à l'environnement : Regards – Recherches – Réflexions*, 4, 207–228.

Blanc, S. and Eymard, C. (2010). Compétences d'un collectif d'école. *Questions Vives*, 5, 43–55.

Boutinet, R. (2005). *Anthropologie du projet*. PUF, Paris.

Charlot, B. (1997). *Du rapport au savoir, éléments pour une théorie*. Anthropos, Paris.

Clanet, C. (1993). *L'interculturel : introduction aux approches interculturelles en éducation et en sciences humaines*. Presses universitaires du Mirail, Toulouse.

Clot, Y. (2007). Psychopathologie du travail clinique [Online]. Available at: www.comprendre-agir.org/images/fichier-dyn/doc/2007/psychopathologie-travail-clinique-activite-yves-clot-chsct.pdf.

Devos-Prieur, O. and Loubet-Gauthier, E. (2002). Incidence des modes de partenariat à l'école élémentaire sur la polyvalence des maîtres et sur les représentations des élèves. *Recherche et formation, les dynamiques identitaires : questions pour la recherche et la formation*, 41, 153–170.

Dubois, M. and Retour, D. (1999). La compétence collective : validation empirique fondées sur les représentations opératoires de travail partagées. *Psychologie du travail et des organisations*, 5, 2(1), 225–243.

Dupont, N. (2010). Les partenariats écoles/institutions culturelles des passages entre cultures juvéniles, cultures artistiques et cultures scolaires ? *Les Sciences de l'éducation – Pour l'Ère nouvelle*, 4(43), 95–121.

Eneau, J. (2016). Autoformation, autonomisation et émancipation. De quelques problématiques en formation d'adultes. *Recherches et éducations*, 21–38.

Fourez, G. (1998). Se représenter et mettre en œuvre l'interdisciplinarité à l'école. *Revue des sciences de l'éducation*, 24(1), 31–50.

Garnier, B. (ed.) (2014). Politiques d'éducation et identités territoriales. In *Carrefours de l'éducation*, Armand Colin, Paris.

Girardot, J.J. (2014). L'observation territoriale coopérative et participative catalyseur de l'intelligence territoriale [Online]. Available at: https://slideplayer.fr/slide/1289485.

Khalali, F., Therriault, G., Bader, B. (2019). Rapport aux savoirs d'élèves du secondaire en lien avec l'environnement et le développement durable en France et au Québec : rapport épistémique et contextualisé au monde. *Éducation et socialisation*, 51 [Online]. Available at: http://journals.openedition.org/edso/5693.

Lange, J.M. (2018). "Education for", territories and positions of teachers: Rupture and tension. In *Evolutions of the Complex Relationship Between Education and Territories*, Barthes, A., Champollion, P., Alpe, Y. (eds). ISTE, London, and John Wiley & Sons, New York.

Le Boterf, G. (2000). *Construire les compétences individuelles et les compétences collectives*. Éditions d'Organisation, Paris.

Levrard, S. and Delalande, J. (2015). Mon école, c'est pas que je la trouve belle, c'est qu'on est ensemble. *Diversité*, 179, 131–135.

Lévy, P. (1994). *L'intelligence collective ; pour une anthropologie du cyberspace*. La Découverte, Paris.

Marcel, J.F., Dupriez, V., Périsset Bagnoud, D., Tardif, M. (2007). *Coordonner, collaborer, coopérer. De nouvelles pratiques enseignantes*. De Boeck Supérieur, Brussels.

Maulini, O. and Perrenoud, P. (2005). La forme scolaire de l'éducation de base : tensions internes et évolutions. In *Les formes de l'éducation : variété et variations*, Maulini, O. and Montandon, C. (eds). De Boeck Supérieur, Brussels.

MEN (1992). Participation d'intervenants extérieurs aux activités d'enseignement dans les écoles maternelles et élémentaires. Report, MEN, Paris, 92–196.

MENESR (2013a). Loi d'orientation et de programmation pour la refondation de l'École de la République [Online]. Available at: http://www.education.gouv.fr/cid2770/le-socle-commun-de-connaissances-et-decompetences.html.

MENESR (2013b). Référentiel des compétences des métiers du professorat et de l'éducation. *B.O.E.N.*, 30.

Mérini, C. (2006). Nature et limites des partenariats éducatifs [Online]. Available at: http://adbdp.web03.b2f-concept.net/spip.php?article421.

Mérini, C. (2007). La dimension organisationnelle du travail enseignant : déterminant de l'ancrage de l'ES dans les apprentissages des élèves. *Premier colloque de la chaire en éducation à la santé du Canada*, Montreal.

Mérini, C. and Bizzoni-Prévieux, C. (2007). Le travail partagé : exploration en formation des maîtres. *Congrès international AREF (Actualité de la Recherche en Éducation et Formation)*, Strasbourg.

Schneuwly, B. and Bronckart, J.-P. (1985). *Vygotsky aujourd'hui*. Delachaux et Niestlé, Lausanne.

Schwartz, Y. and Durrive, L. (2003). *Travail et ergologie, entretiens sur l'activité humaine*. Octarès Editions, Toulouse.

Thélot, C. (2004). Pour la réussite de tous les élèves : rapport de la commission du débat national sur l'avenir de l'école présidée par Claude Thélot. Report, La Documentation française, Paris.

Thomazet, S. and Mérini, C. (2014). Le travail collectif, outil d'une école inclusive ? *Questions Vives*, 21 [Online]. Available at: http://questionsvives.revues.org/1509.

Vincent, G. (1994). *L'éducation prisonnière de la forme scolaire ? Scolarisation et socialisation dans les sociétés industrielles*. Presses universitaires de Lyon, Lyon.

Wittorski, R. (1997). *Analyse du travail et production de compétences collectives*. L'Harmattan, Paris.

Zarifian, P. (1999). *Objectif compétence. Pour une nouvelle logique*. Liaisons, Paris.

Zay, D. (1994). Enseignants et partenaires de l'école. Démarches et instruments pour travailler ensemble. *Recherche et Formation*, 17, 166–169.

9

The Instrumentalization of Education in Sustainable Development at the Service of Tourism: The Case of the Arganeraie

9.1. Introduction

The relationship between tourism and environmental resources, whether natural or cultural, material or immaterial, is often considered as the cause for multiple conflicts and incompatibilities. The negative repercussions of the movement of hundreds of millions of travelers are denounced every year. About 30 years ago, Georges Gazes emphasized the seemingly contradictory, but simultaneously complementary, "double logic" which inspires any tourist enterprise: on the one hand, the need to valorize the local patrimony, to make it profitable by opening it widely to remunerative visits, and, on the other hand, the need to safeguard this environmental capital, even to limit its access to make it last longer (Cazes 1992, pp. 124–125).

This chapter attempts to overcome these conflicts, with mixed success, by analyzing the policy enforced at the Arganeraie biosphere reserve (ABR) in Morocco based on its educational dimension and its partially unexplored potential.

9.2. Environmental crisis and inflation of alternative tourism

It is no longer necessary to demonstrate that tourism is a fundamental economic sector for a large number of countries. The 2020 health crisis provided additional evidence of the fall in the GDP of many countries whose economy is particularly

---

Chapter written by Salma ITSMAÏL and Bruno GARNIER.

dependent on tourism. In early July 2020, the United Nations Conference on Trade and Development announced that the financial loss could reach USD 2.2 trillion, or 2.8% of the global GDP, if the suspension of international tourism lasted eight months, matching the expected decline in tourism predicted by the World Tourism Organization. According to Pamela Coke-Hamilton[1], "for many countries, like Small Islands and Developing States (SIDS), a collapse in tourism means a collapse in their development prospects. It is not something we can afford" (2020). In the most seriously affected countries, which are both developed and developing countries with high tourism potential, or in certain subnational regions, such as Corsica or the Balearic Islands, tourism is an essential factor and a driver of economic and social development. It is considered as an important source of foreign exchange and a first-class job creator. For decades, the touristic development of natural resources has constituted a powerful vector for territorial transformation. With a potential influence on all economic sectors, tourism affects agriculture, craftsmanship, transport and trade, insofar as rural handicrafts are mainly consumed by tourists.

However, the limits of the tourist sector have only recently been taken into account, whereas the frequency of trips has intensified and the duration of stays has shortened (Afeissa 2007). Internationally renowned places have to endure exceptional concentrations of visitors for brief periods of time (such as Saint Mark's Square in Venice) and the craving for the foreign extends to the most remote corners of the world, such as the 4,000 islands on the Mekong river in Laotian territory (Preuil 2015).

The depletion of certain natural resources particularly attractive for tourists was a first alarm signal, while the consequences of a practice exclusively oriented to the economy proved to be disastrous, particularly in relation to the production of greenhouse gases and atmospheric pollution by air transport, in a period of increasingly worrying global warming. The belated awareness of the tension between tourism and the preservation of the natural and cultural environment and of biodiversity stimulated the search for new strategies to sustainably ensure the significant inflows of currency from the tourism sector. Radical changes began to affect the management of tourist sites, promoting values such as respect for the environment and the preservation of patrimonial identity.

In recent years, under the effect of several converging factors (awareness of issues related to biodiversity, the willingness of receiving states to protect their patrimony), new forms of alternative tourism have developed, including community, solidarity and ecotourism. These have been presented as a means of reconciling the

---

1 Head of International Trade at the United Nations Conference on Trade and Development (UNCTAD).

economic development of countries, the well-being of local populations and the protection of the environment and associated natural resources. In a context of increased media coverage of issues related to biodiversity and its threats, tourism (and particularly ecotourism and nature tourism) is often presented as a means of reconciling conservation with development (Orams 1995; Breton 2001; Lequin 2001, 2002; Caccomo 2007; Buckley 2008, 2010; Buckley et al. 2008).

This new strategy can adopt several manifestations, among which the main accepted definitions have been discussed (Hall 2003; Laliberté 2005): the term "ecotourism" refers to forms of tourism set in a natural environment for the purpose of learning; "fair tourism" is generally associated with north–south relations and is inspired by fair trade principles, ensuring that local communities are involved in the tourist service and benefit from its economic effects, with the goal of improving the living conditions of the inhabitants of the territories involved.

While the term "solidarity tourism" is close to the previous one, it emphasizes the links between people – in this case between visitors and hosts – through the notion of solidarity. In this case, travelers contribute to the improvement of the living conditions of the communities visited. "Responsible tourism", also known as "ethical tourism", refers to the tourist's way of traveling and social conscience, whereas "social tourism" promotes the right to holidays and accessibility to tourism for all the sectors in the population, with a special emphasis on minority groups.

9.3. Tourism and sustainable development

We should bear in mind that, in the field, it is almost always hybrid forms of these different definitions that emerge. Thus, the authors of this chapter have chosen the term "sustainable tourism" as the most suitable for the particular case of the ABR because it associates "tourism" with "sustainable development" (François 2004). On the one hand, "sustainable tourism" conveys the idea that "tourist activities must be conducted in harmony with the specificities and traditions of the host regions and countries" (OMT 1999), while, on the other hand, the notion takes into account the impact on the environment insofar as it helps "people to lead healthy and productive lives in harmony with nature" (Shackelford 2002).

In this new form, sustainable tourism tends to become a component of sustainable development, granting tourism development while preserving the chances for the future. At the same time, sustainable tourism tends to become a form of non-formal education in sustainable development. Sustainable tourism has to be educational tourism or else it shall not be, we may be tempted to say. Several factors have to be taken into consideration for its implementation. The tourist sector mainly depends on sustainable management to cope with the harmful effects of mass tourism. The sustainability of this sector is based on the preservation of nature, landscapes and the conservation of a cultural identity. The local population, committed to the conservation of the territory and the patrimony, plays a fundamental role in this. The target audience for the education in sustainable development through tourism is twofold: it includes visitors and hosts.

From this perspective, the relationship between tourism and sustainable development is inseparable and enhances the positive effects of the tourist sector by means of an ecological strategy. Even though the debate on sustainable development is often related to ecological requirements, development sustainability implies taking into account the economic, political, cultural and social spheres. In short, sustainable tourism shows respect for cultural identity, the physical environment and social cohesion.

The sustainable development of tourism relies on the adoption of a quality educational policy, which requires a change in the behavior of suppliers and visitors.

9.4. Sustainable tourism and patrimony: educational issues

At first sight, the notion of sustainable tourism, associated with that of sustainable development, seems far from the notion of patrimony. Historically, it was the economic problems related to the depletion of natural resources that motivated reflection in favor of sustainable tourism, a concept derived from the notion of economically sustainable development, based on the preservation of natural resources. Environmental problems have been the subject of several studies since 1972. Indeed, the Maedows report mentioned the jeopardy of uncontrolled economic growth over natural resources, foreseeing their depletion. During the same period, the United Nations Conference for Development stressed the importance of environmental protection for the maintenance of economic development. The purpose of these two events was to raise awareness on the need to integrate the environment into the decision-making process, which led to the implementation of development strategies based on environmental protection.

Later on, the Brundtland report was issued by the World Commission on Environment and Development in 1987, which described the notion of “sustainable development”. The report presented the protection and improvement of the environment as a global priority, requiring an urgent change in economic strategies. This report was considered as a baseline for the United Nations Conference on Environment and Development held in Rio. In fact, in 1992, the “Earth Summit”, organized by the United Nations Conference on Environment and Development (UNCED) made it possible to concretize the ideas presented in the Brundtland report through the signing of international conventions on climate change and biological diversity and by setting up various action programs.

However, while the “Earth Summit” insisted on the importance of integrating the environment in the decision-making process, the third pillar of sustainable development, namely humans and the local population, seemed to have been forgotten. It was not until the Vienna Summit in 1994, which supplemented the UNCED, that the priority role of humans in achieving sustainable development was asserted. This involved introducing ethical responsibility through education in resource sharing. However, a major issue emerged in this sense: it became essential to educate the present generations, laying down the rules for equitable sharing. This goal was then associated with the definition of patrimony insofar as it constitutes the “global patrimony”.

Ultimately, sustainable development and patrimony meet in at least three points.

First, the two notions are based on the idea of solidarity and a sharing spirit: patrimony is transmitted from one generation to another, this transmission process only being possible through the adoption of a sustainable development strategy, so as to meet the needs of the present generation without compromising the possibility for future generations to meet theirs.

Second, sustainable development requires understanding the territory, identifying the area’s specificities before the implementation of a sustainable development strategy. The same happens with patrimony, which is identified by its specificity; its identity is specific to the territory.

Third, the role of the local population is omnipresent in both notions. Based on Egyptian culture, Cheikh Anta Diop identified three factors rooting the identity of a person living in a territory: historical, linguistic and psychological factors. The historical factor unites the disparate elements of a people into a whole, creating a sense of continuity shared by the entire collectivity. The historical feeling allows the people to feel that they are something greater than a population. The linguistic factor concerns language and all the means of communication that people use to translate their emotions. Finally, the psychological factor brings together the psychic invariants that keep the construction of the person on a territory intact (Diop 1981). The three above-mentioned factors cannot exist unless they are transmitted through education, that is, by becoming common patrimony and being recognized as such in the building of a collective identity, as well as the identities of the individuals: all different, but perceiving themselves as similar through the feeling of the shared patrimony transmitted to them (Garnier 2014).

The existence and continuity of the cultural patrimony depend on the forms of education available to the population bearing the identity and possessing specific information or know-how. It is not limited to formal education because formal education, as provided in educational institutions, the fruit of the long patrimony of the school form (Vincent 1980), is cut off from the context of everyday life and the social and political environment of learners (Garnier 2018). Non-formal education, in contrast, is halfway between informal education – which is not conscious and works by osmosis with the environment, such as the learning of oral language in young children – and formal education.

Non-formal education is a form of "contextualized" education (Rogers 2004) insofar as it transmits knowledge, skills and life skills based on experience without being subject to institutionalized programs outside the context of daily life and production activities. Non-formal education can be transmitted in the family, in the village, by an elder, by imitation, by hearsay, by seeing and also within the framework of voluntary educational programs, in particular those including sustainable tourism and those where the school can find inspiration, especially in the context of "education for" (Barthes and Alpe 2012). At a time when patrimony has become as fragile as biodiversity, non-formal education seems essential to the survival of a culture which runs the risk of no longer being transmitted via the intergenerational relationship due to the social upheavals of modern life, which hinder a transmission we are tempted to qualify as "natural transmission". Sustainable development can in no way meet its challenges without the non-formal education of the local population living in the territory. Collective behavior is a movement to be maintained because it threatens to come to a definitive halt, no longer sufficient in itself.

Therefore, it appears that the patrimony and sustainable development are closely related, and it becomes possible to state that the continuity and existence of patrimony depend on the application of a sustainable development strategy comprising a strong educational dimension. It is also an issue for tourism sustainability. While a tourist may be initially attracted by the quality of the local patrimony, the damage caused in unprotected historical sites (or where environmental concerns are neglected) makes local and international tourists flee away from such places. Even before the official appearance of the notion of sustainable development, R.W. Butler had concluded that tourist attractions are non-renewable, and he considered environmental conservation, protection and preservation necessary to ensure a sustainable tourist life cycle. He wrote, "the development of tourist locations must be kept within predetermined capacity limits, and their competitive potential maintained for a long period" (Butler 1980).

Sustainable development makes it possible to valorize patrimony by encouraging the visitor and the hosts to preserve the country's identity and culture, as "a new tourist product" centered around local patrimony and biodiversity. However, as has been shown, sustainable development has become a major concern for maintaining economic growth. Tourism, which represents a major sector in each country, cannot ignore the protection of the natural patrimony.

9.5. Towards a "sustainable strategy"

The concept of sustainable development, which first appeared in 1987 in the already mentioned report by the United Nations World Commission on Environment and Development, was taken up in 1991 in the report *Caring for the Earth* published by the United Nations Environment Programme (UNEP) and the International Union for the Conservation of Nature (IUCN), both of which highlight the main themes of sustainable tourism, stating that humanity must take into consideration the Earth's capacity limit by adopting a lifestyle and development model respecting nature (Munro 1991). This report described sustainable development as the path for improving life while conserving the diversity planet Earth offers us. Shortly after, the World Tourism Organization (WTO) adopted the concept of sustainable development in the tourism sector. Since 1993, several of the WTO's publications have incorporated the expression "sustainable tourism development". In the guide to sustainable tourism development, the WTO refers to the optimal use of resources which are a key element of tourism development, preserving essential ecological processes and helping to conserve natural patrimony and biodiversity.

Among the actions launched by the WTO to ensure sustainable tourism, we can mention the creation of a world charter for sustainable tourism. In 1995, the WTO, UNESCO, UNEP and the European Commission drafted a charter for sustainable tourism enabling the dissemination and implementation of sustainable tourism at a global scale: 10 articles regulate sustainable tourism throughout the world. Mention can also be made regarding the agenda for the implementation of sustainable tourism by means of a program entitled "The Promotion of Environmental Protection and Sustainable Development in Tourism". The strategic goals of Agenda 21 are the prevention of the negative impact of mass tourism on territories and the environment, the maintenance of sustainable local development and the adoption of responsible tourism.

The WTO has inspired several labels to encourage sustainable tourism, such as the "Gîte Panda", which labels accommodation located in a natural environment favoring sustainable tourism, and "la Clef Verte", which rewards accommodation raising awareness about the environment, in connection with "the foundation for environmental education". These labels encourage the dissemination of environmental education for employees and customers.

It should be noted that the actions implemented by the WTO only popularize the concept of sustainable tourism but cannot claim a form of sustainable tourism adaptable to every place. However, the deployment of a responsible tourism strategy presupposes a fine adaptation to the place's reality, taking into account the territory‘s identity, culture and patrimony.

The theory of *symbolic sites of belonging* considers these as “a set of collective landmarks, bearing the marks of the site's ethics and history” (Zaoual 2006, p. 19). The acknowledgment of the territory‘s diversity leads to a better comprehension of the site's actors, and, on this basis, the managers of such places can model a specific and effective strategy to ensure the local development. In a given territory, the “actors act according to their rationality and modes of coordination, which bear the mark of their *symbolic site of belonging*”. The site is considered as “a cognitive space which prompts adjustments between the individual and the local community” (Zaoual 2006, p. 18).

We can therefore rule out the hypothesis on the existence of uniform touristic development and should rather assume the existence of specific development, taking into consideration a territory's diversity. When adapting the notion of “symbolic sites of belonging” to sustainable tourism, it appears that one of the basic principles of this theory is to acknowledge the territory's identity, the common values of the local population, as well as the modes of patrimony transmission from generation to generation. It thus becomes evident that sustainable development cannot be implemented without the integration of the local population in the management strategy. It is impossible to achieve sustainable tourism without involving the population in the process of change.

The influence of population on the sustainability of the tourism sector should not be overlooked. This reinforces the idea that education in sustainable development is necessary and needs to be put in place to achieve the goal of local economic development.

9.6. The Moroccan situation: a sustainable tourism policy in the ABR?

As with several other countries, Morocco is committed to integrating the principles of sustainability in all sectors, including tourism. Morocco's commitment was officially announced in April 2017 in the government's declaration before parliament, expressing the will to align itself with global efforts to reach the sustainable development goals. Since then, several reference systems have emerged, among which the most important are: the National Charter for the Environment and Sustainable Development, the Green Morocco Plan, the National Human

Development Initiative and the National Energy Strategy (NES). We here focus on the ABR.

The concept of a "biosphere reserve" aims to find a balance between the conservation of biological diversity, the search for economic and social sustainable development and the maintenance of a territory's own cultural identity. In December 1998, the Arganeraie (centered on the argan tree and its ecological system) was recognized by UNESCO as a "biosphere reserve". The area covers a surface of 2.5 million hectares and is home to a population of nearly 2.6 million inhabitants in the provinces and prefectures of Agadir Ida Outanane, Inezgane Ait Melloul, Chtouka Ait Baha, Tiznit, Taroudant, Sidi Ifni and Essaouira. This reserve was created around a forest species endemic to Morocco known as the argan tree, whose economic value is important for the region and for the country. The socio-economic development of the area is strongly related to the various products and multiple uses of the argan tree.

The argan tree ecosystem has suffered great degradation, which has made its regeneration almost non-existent. Since the labeling of the argan tree territory, several projects have been established, such as, for example, the "Circular Economy Approach for the Conservation of Agrobiodiversity in the Souss-Massa Draa Region in Morocco", a project set up by the French Agricultural Development Agency (ADA) in partnership with other management stakeholders of the Souss-Massa region. One of this project's goals is the promotion of local products and the marketing of natural patrimony for sustainable tourism.

A few years ago, Morocco shifted its development strategy towards the tourism sector, defined as a powerful locomotive for the country's development, to create jobs and generate foreign currency. Tourism was classified as a priority sector, receiving particular attention from the Moroccan State. Since the royal instructions issued in 2001, Morocco has been offering new tourist products with a multicultural vocation, and we are witnessing the birth of a new form of tourism, rural tourism, which has taken on a large scale, valorizing the cultural patrimony and integrating local populations in the development process. This form of tourism has been supported by local development associations, international NGOs and development agencies.

The rural development network aims to highlight the economic and sociocultural benefits of rural tourism in Morocco through the promotion of tourism in the country's natural areas. Among the several projects deployed, we can mention the action plan for the development of ecotourism in protected areas.

The ABR has benefited from particular attention, given the area's specificity and its exclusivity to Morocco. The ABR offers an exceptional tourist product thanks to the Atlas Mountains and the local Berber population, whose culture alone represents a treasure to be discovered, in particular, its craftsmanship and culinary traditions. This tourist offer is mainly based on the cultural dimension, the historical sites, the natural patrimony and the stunning mountains, which provide an attractive backdrop. The city of Agadir, capital of the argan tree, is one of the most visited tourist destinations in the country. However, it is seaside tourism that dominates to the detriment of the "argan zones", which can offer new forms of tourism (ecotourism, rural tourism, cultural tourism or mountain tourism).

The ABR represents a space rich in biodiversity, as well as an architecturally significant patrimony. The valleys, the oases, the collective granaries, the traces of dinosaurs, the engravings and the sacred argan tree are all symbols of an ancient sedentary civilization that valorize the territory and promote sustainable tourism. Hence, the region was awarded the "protected area" and the "biosphere reserve" statuses by UNESCO. Beyond its status as a biosphere reserve granted by UNESCO, the ABR has been able to make people talk about it: in 2017, it received the "sustainable forest prize" from the Spanish tourism magazine *Aire Libre* in recognition for the actions implemented for the conservation of biodiversity, the valorization of patrimony and sustainable development.

This prize was the fruit of the "Tourist Host Country of the Argan Tree" project, which sealed a partnership agreement in sustainable tourism development in some provinces of the ABR. The agreement was signed in 2009 by the Minister of Tourism and Handicrafts, the High Commissioner for Water and Forests and the Fight Against Desertification, the Secretary of State in charge of water and the environment, the president of the regional council of Souss-Massa-Drâa and the president of the General Council of Tourism in Agadir.

This agreement gave added value to the kingdom's tourist offer through the involvement of all partners in the protection of the region's natural and cultural assets, as well as by launching programs for the development of appropriate road infrastructure, the construction of light reception facilities and the development of the bed and breakfast policy favoring responsible tourism.

In 2015, a second partnership agreement for sustainable tourism development in natural areas was signed by the Minister of Tourism and the High Commissioner for Water and Forests and the Fight Against Desertification, aimed at developing an integrated and sustainable tourist offer in Morocco's protected areas. This agreement made it possible to develop tourist activities in natural areas, to enrich the offerings and to promote tourism while respecting the region's ecosystem. This resulted in Morocco being included among the countries promoting sustainable development by offering nature, ecological and rural tourism to its visitors, encouraging sustainable tourism as part of its management strategy.

Other support actions were planned under the agreement to achieve the sustainable tourism goal, among which we can mention training tourism partners regarding the concept of sustainable development, education in good practices for the preservation of the ecosystem, the elaboration of sustainability standards and the development of tourism in protected areas.

In 2019, the standard from the Arganeraie Biosphere Reserve Ecolabel Moroccan Association (ABR EMA) specifically started regulating the argan tree biosphere reserve territory via the implementation of an ABR ecolabel certification system. This standard aims to regulate all the activities of this zone (consumption, production, rural tourism, etc.). The units displaying the ABR Ecolabel must respect a positive economic, social and environmental strategy that targets the goals specified by the 2030 United Nations sustainable development agenda. The standard is designed to respond to the problem of the regeneration of the argan tree ecosystem through the introduction of payments for ecosystem services. This repository administers several activities in the ABR territory, including the tourism sector.

9.7. A cultural as well as a natural patrimony item: the argan tree

The ABR occupies 7% of Morocco's forest surface. The local population comprises mostly Amazighs. Nearly 2 million people are directly affected by the exploitation of the argan tree (Faouzi and Martin 2014). For the region's local population, the argan tree represents a precious cultural symbol and constitutes the sociocultural identity of the area. From a purely economic point of view, the argan tree represents the population's main source of income. In contrast, from an ecological point of view, the argan forest is a natural barrier against the advance of the desert because it enables the protection of the soil from erosion, as well as that of the territory's fauna and flora. The ABR constitutes an ecological balance for the region. However, unfortunately, the argan tree has lost half of its surface area within a century (Tarrier and Benzyane 2003).

The infrastructure in the biosphere reserve is well developed compared to other regions, while the road network grants access to the majority of municipalities in the area. Every year, millions of visitors travel to the ABR via road networks and airports (Al Massira Agadir airport and Essaouira-Mogador airport). The airport and road infrastructure have made it possible to develop a rural tourism product by facilitating access for tourists and travel agencies organizing excursions to the ABR.

9.8. Between reality and opportunism: the instrumentalization of sustainable development

Despite the efforts made to promote the recognition of the patrimony to be preserved, the number of tourist establishments respecting sustainability guidelines is minimal. According to the records, the ABR has 136 accommodation establishments classified as hotels, lodges or guest houses, with little concern for the respect for and dissemination of sustainable tourism principles (Aboutayeb 2015). The city of Agadir attracts a large number of visitors every year. This makes travel agencies offer several excursions and tourist circuits to the ABR territory, to such an extent that Agadir agencies are responsible for the most frequent visits to the area. The main tourist excursions are Chtouka-Ait Baha, Essaouira, Immouzzar-Ida Ou Tanane, Massa and the National Park, Tafraoute, Taghazout and Taroudant. These travel agencies also offer sports trips (hiking, mountain biking, fishing, hunting). Unfortunately this kind of structure is still not managed by any administration, and the offers for visiting the ABR are not strictly regulated. Despite the policies announced by Moroccan institutions in favor of sustainable tourism, some structures are still not regulated, which hinders the image of the sustainability of the ABR territory.

In order to deepen this analysis, it is necessary to get to know the representations of the visitors to the area in terms of sustainable development and natural and cultural patrimony. To do this, as part of her doctoral research, Salma Itsmaïl undertook a study of the perceptions of Moroccan and French tourists (the main tourist market for Morocco) regarding the ABR destination.

In order to depict the reality of sustainable tourism in the ABR, the researcher strove to find out how tourists perceive sustainable tourism in the ABR by means of the excursions made to the reserve. A survey was conducted among Moroccan as well as French tourists who had visited Morocco at least once. The purpose was to obtain a truthful picture of sustainable tourism in the ABR. First, it was necessary to determine the percentage of tourists who had an idea of the concept of sustainable tourism; second, the percentage of tourists who had already heard of the ABR; and finally the percentage of tourists who had already made a "sustainable" trip to the argan tree biosphere reserve. Among the sampling methods, "sampling by convenience" was chosen. The survey was conducted among the tourists who agreed to answer to the questionnaire. A snowballing process was also used to reach a large number of study partipicants (Gauthier and Bourgeois 2016). A total of 172 responses were collected from French tourists and 198 responses from Moroccan tourists. It turned out that the results obtained in the two samples were very similar for all the questions asked.

The first topic concerned knowledge of the concept of sustainable tourism: 62% of Moroccan tourists and 58% of French tourists said they had already heard of it. Tourists familiar with the principle of sustainable tourism were largely among the 26-to-49-year-old group, part of the active population.

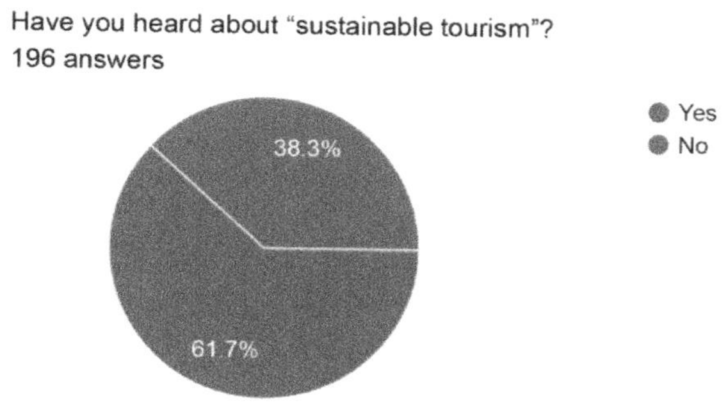

Figure 9.1. *Moroccan sampling result: knowledge of the concept of sustainable tourism. For a color version of this figure, see www.iste.co.uk/barthes/biosphere1.zip*

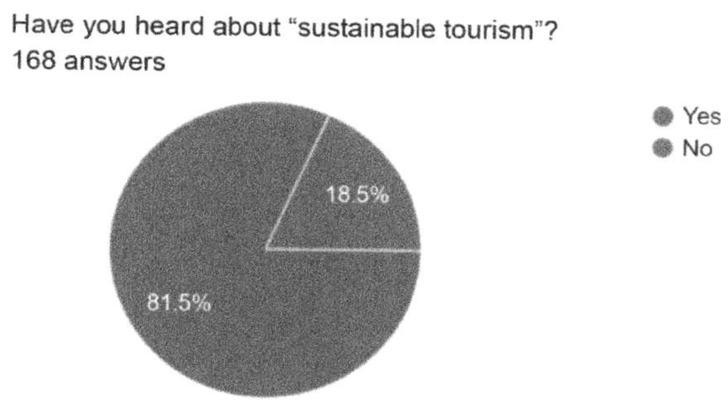

Figure 9.2. *French sampling result: knowledge of the concept of sustainable tourism. For a color version of this figure, see www.iste.co.uk/barthes/biosphere1.zip*

In order to define the territory studied, tourists were asked whether they had ever come across sustainable tourism offers in Morocco: 42% of Moroccan tourists and 11% of French tourists said they had already heard of sustainable tourism in Morocco. Their sources of information were websites in the first place and TV shows. Obviously, websites are increasingly becoming a source of communication, playing a fundamental role in the transmission of information.

A first reading of the results brings to light the lack of information concerning sustainable tourism offers in Morocco because the media coverage of this type of tourist offer is negligible.

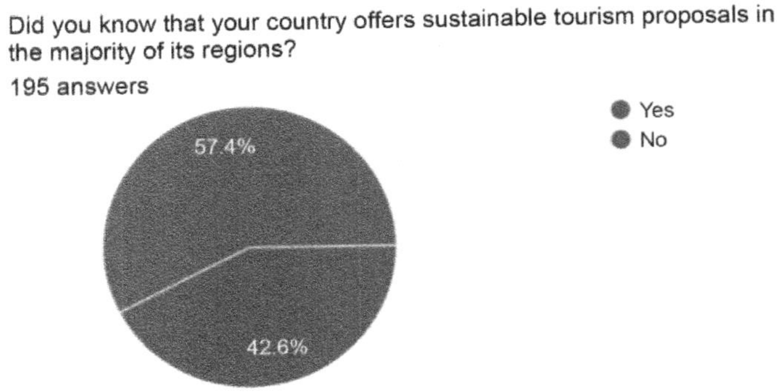

Figure 9.3. *Moroccan sampling result: information sources for the notion of sustainable tourism in Morocco. For a color version of this figure, see www.iste.co.uk/barthes/biosphere1.zip*

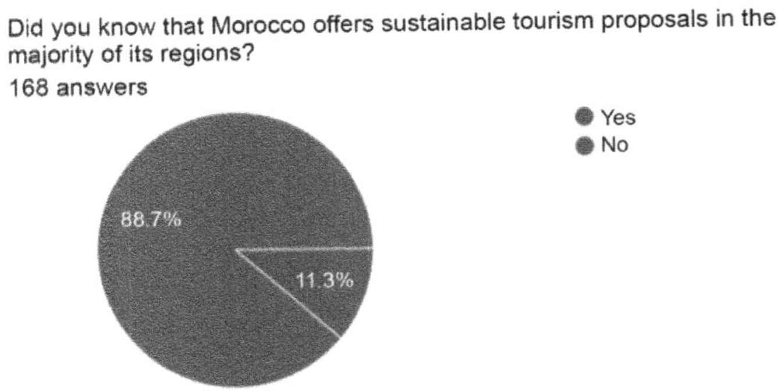

Figure 9.4. *French sampling result: information sources for the notion of sustainable tourism in Morocco. For a color version of this figure, see www.iste.co.uk/barthes/biosphere1.zip*

Pushing the reflection further, the tourists were questioned about their knowledge of the "biosphere reserve" label before being questioned on the existence of the "Arganeraie biosphere reserve". It turned out that only 22% of Moroccan tourists (against 6% of French tourists) reported having heard about biosphere reserves. Despite the label's history, it was largely unknown to the majority of the subjects questioned, which leads to the hypothesis that the awareness-raising process put in place by the reference countries has not been effective so far.

As for the Arganeraie biosphere reserve in Morocco, 29% of Moroccan tourists and 10% of French tourists reported that they had already heard of this reserve. The lack of information on the ABR is immediately explained by the obscurity of the label itself.

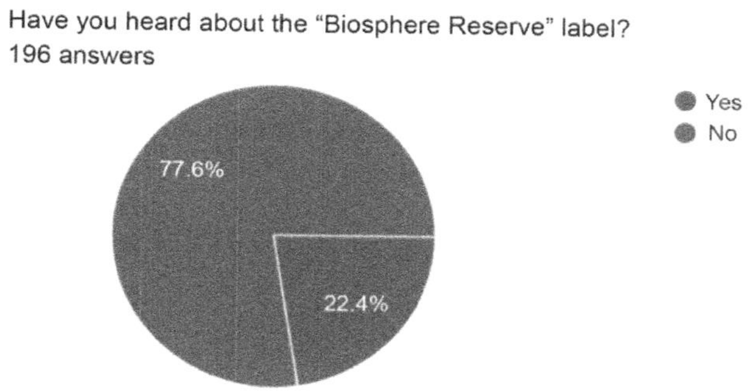

Figure 9.5. *Moroccan sampling result: knowledge of the "Biosphere Reserve" label. For a color version of this figure, see www.iste.co.uk/barthes/biosphere1.zip*

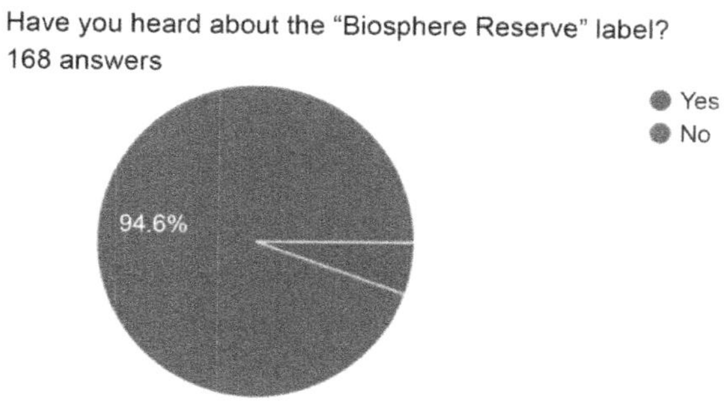

Figure 9.6. *French sampling result: knowledge of the "Biosphere Reserve" label. For a color version of this figure, see www.iste.co.uk/barthes/biosphere1.zip*

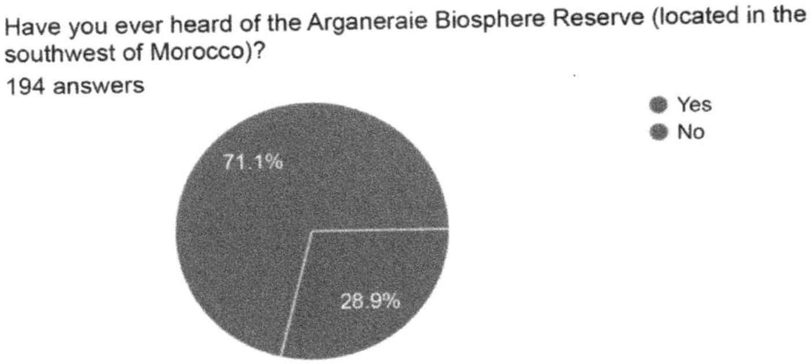

Figure 9.7. *Moroccan sampling result: knowledge of the Arganeraie "Biosphere Reserve" label. For a color version of this figure, see www.iste.co.uk/barthes/biosphere1.zip*

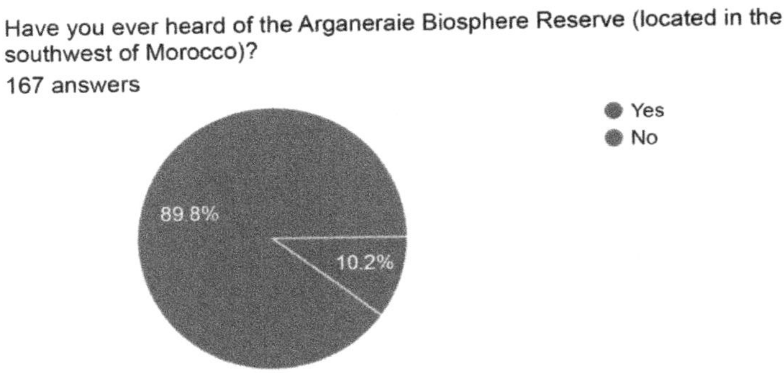

Figure 9.8. *French sampling result: knowledge of the Arganeraie "Biosphere Reserve" label. For a color version of this figure, see www.iste.co.uk/barthes/biosphere1.zip*

In order to evaluate the real aspects of the question in the field and to break the barrier of the absence of information on the zone's label, we questioned the tourists whether they had opted for a "responsible trip" to one of the areas of the ABR: 72% of Moroccan tourists and 87% of French tourists declared they had never opted for this type of tourism to the ABR, pointing to the lack of information on the subject (69% of Moroccan tourists and 92% of French tourists criticized the total absence of communication on the part of the Moroccan state regarding sustainable tourism in the ABR). However, the agencies' offer was not considered particularly "interesting" in their opinion (14% of Moroccan tourists and 5% of French tourists deemed the offers of travel agencies uninteresting).

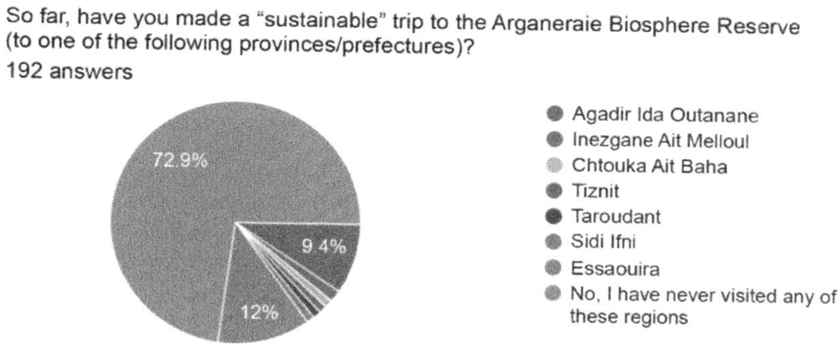

Figure 9.9. *Moroccan sampling result: experience of a "sustainable" trip to the argan tree reserve. For a color version of this figure, see www.iste.co.uk/barthes/biosphere1.zip*

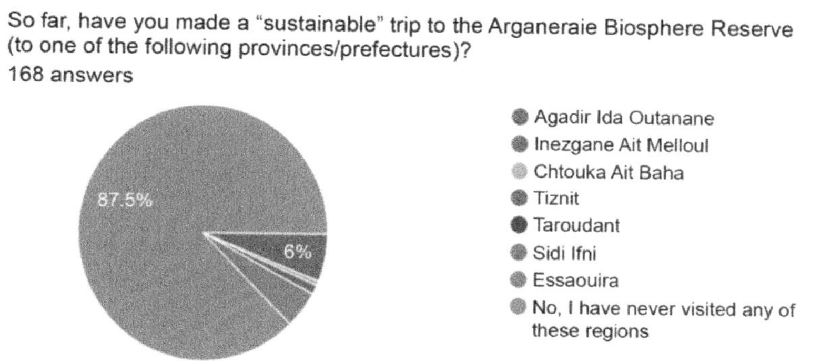

Figure 9.10. *French sampling result: experience of a "sustainable" trip to the argan tree reserve. For a color version of this figure, see www.iste.co.uk/barthes/biosphere1.zip*

The results obtained by the survey show a real gap between the efforts made by the administrations and the reality in the field. It seems urgent to review the strategy put in place to promote sustainable tourism within the ABR. While the commitment of sustainable development decision-makers is present, the results evidenced are minimal. It became patent that over 40% of the population had never heard of sustainable tourism. This figure proves the lack of information and media coverage in the field of sustainable development, as well as the limited offers of travel agencies in terms of responsible stays. However, the ABR has the potential to become a leading destination for sustainable tourism, likely to promote economic, social and cultural development in the area in accordance with sustainable development principles.

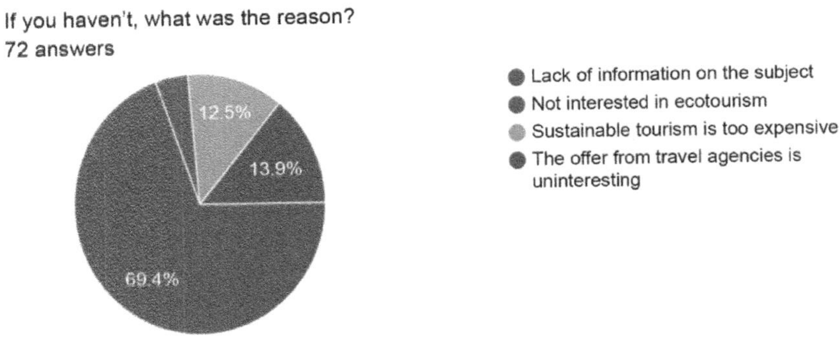

Figure 9.11. *Moroccan sampling result: reasons for not choosing a "sustainable" trip to the argan tree reserve. For a color version of this figure, see www.iste.co.uk/barthes/biosphere1.zip*

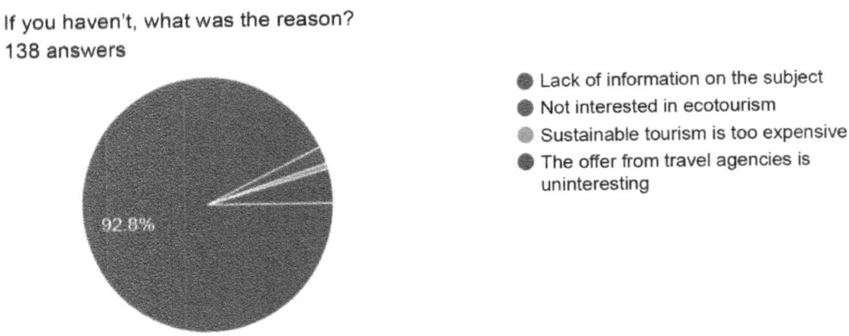

Figure 9.12. *French sampling result: reasons for not choosing a "sustainable" trip to the argan tree reserve. For a color version of this figure, see www.iste.co.uk/barthes/biosphere1.zip*

This reductive vision is directly related to the services offered by travel agencies, the organizational processes and the resources used by tourism operators.

9.9. Education: the missing vector for sustainable tourism

The results of the above-mentioned surveys stress the importance of proper education in sustainable development, both for the local population of the region and for the tourist actors who intervene on the territory. Three axes can be proposed for the deployment of a sustainable tourism strategy centered on sustainable development: the implementation of specifications for tourist products intended for the reserve, environmental education and awareness to ensure the preservation of natural patrimony and the promotion of "responsible" tourism products in relation to the ABR. Field research has shown that we have not yet found a way to go beyond the "instrumentalization stage in education for sustainable development".

This instrumentalization is reflected in the fact that, for tourism operators, the various labels awarded to sites to be preserved have no other interest than to attract local or foreign customers to traditional tourist products, free of educational and environmental concerns. To propitiate a real paradigm shift, it would still be necessary to highlight the reality of sustainable tourism as an opportunity to educate citizens from the host country, as well as tourists, on environment-friendly practices.

In 1972, the Vienna Charter promulgated by the General Assembly of the International Office of Social Tourism presented tourism as an inalienable right that enhances the development of the individual and brings to light the deep bond between peoples. They stated that "tourism should be conceived as a privileged means for human maturation and for raising human awareness on the profound unity of humanity" (BITS 1972). According to this charter, tourism can be considered as a real factor of the transmission of information and several researchers have shown the impact of tourism in educational matters. In 1962, the sociologist Joffre Dumazedier was the first to mention the three Ds that can be found in tourist activities: *délaissement* (relaxation), *divertissement* (fun) and *développement* (development). He wrote: "leisure is a set of occupations individuals can devote themselves to out of their own free will, either to rest, to amuse themselves, to gather information, to follow disinterested training, engage in voluntary participation or foster their free creative capacity, after having freed themselves from their professional, family and social obligations" (Dumazedier 1962, p. 29). The least that can be said is that tourism as practiced by commercial actors operating in the ABR shows little concern for non-formal education in sustainable development nor even for the true knowledge of patrimony transmission modes that inspires a feeling of belonging to a human community on this territory.

This is all the more regrettable as tourism has the power to convey a cultural image and to educate individuals on various topics. Not only does this idea apply to the cultural and architectural patrimony, but also to the natural patrimony. The natural landscape contributes to sharing environmental knowledge and fostering respect for the global ecosystem. Tourism can be oriented – without this being its main goal – towards non-formal education, which is why it is important to train tourism actors in sustainable development. It is necessary for each participant to be involved in the transmission of education respecting the environment.

Moreover, the ABR, and more specifically, the Souss Massa National Park, has set up codes of ethics to be respected by visitors concerning wild animals, their natural habitat and respect for the flora. Tourism is theoretically defined as sustainable and educational tourism. The idea of integrating environmental education into the tourist circuit, favoring the transmission of information on the region‘s fauna and flora, complies with the international strategy for educational/didactic programs, according to which “tourism has a local and national organization and influence, and the actions must also match these levels. Tourism guidelines should be formulated with a view to conserving the resources from the region’s natural and cultural environment” (UNESCO and UNEP 1987). This program calls for the implementation of a model for environmental education and information, raising awareness on the environment’s importance. It aims to engage people around the conservation of natural patrimony and to make them aware of the importance of the environment for the development of the zone.

Several other activities offered in the ABR relate to environmental education, and this type of offer does not fail to attract tourists by stimulating their curiosity and their desire to learn. However, it would be necessary to appeal to all parties, namely the local population, travel agencies and administrations, to stimulate the synergy between patrimony and the development of learning tourism.

The integration of the local population, in particular, is essential for any tourist activity aimed at highlighting the regional patrimony and exploiting learning tourism. The indigenous people would take on the role of “patrimony guardians” as guides and bearers of information. They would feel a sense of pride by sharing their knowledge with visitors. The indigenous people would play a key role not only in the preservation of the natural patrimony but also in the development of learning tourism.

Some ABR volunteer organizations are already committed to this through the training of local guides, by raising awareness on pollution and through the conservation of the regional fauna and flora and environmental education. From this perspective, during the Moscow congress in 1987, a precursory researcher, Rozanov Boris, evoked the fact that education changes orientation and helps reconsider the way in which the environment is approached. He asserted that the problems related to the environment have to be solved directly in the field by engaging the indigenous population. Within the framework of the symposium One Region, One Planet in 1991, Roberge Charles evoked the importance of the preservation of biodiversity through the integration of populations, which are an integral part of the ecosystem.

According to the Seville strategy, it is necessary to strengthen training and education on biosphere reserves through the implementation of popularization, information and education programs promoting the conservation and sustainable use of natural resources. A reading of the Seville strategy shows that environmental education and education in sustainable development are omnipresent in a repetitive way.

9.10. Conclusion

There is still a long way to go before most, if not all, tourism businesses become "legal persons", that is, not merely companies generating short-term profits and in danger of disappearing due to the depletion of natural resources and the consequences of global warming on the restriction of transport. The aim is for tourism actors to become socially responsible and for all travelers to be respectful of the environment visited. The process is in motion. On the ABR territory. there are many educational associations and establishments, among which the most important are the Souss Massa National Park, the network of associations of the ABR and the southern foundation on the initiative of thematic museums.

However, these actors seem isolated and to lack coordination and synergy. The impact of their actions on the representations of the public concerned by tourism products is still minimal. In certain aspects, the government's actions in favor of patrimony seem to respond to the immediate economic imperatives of travel agencies. The training of all the actors is a necessity if we want sustainable tourism and patrimony-based education not to fall on deaf ears and not to be instrumentalized.

9.11. References

Aboutayeb, H. (2015). *Tourisme durable dans la réserve de biosphère de l'Arganeraie au Maroc. Plaidoyer pour l'implémentation d'une stratégie de développement touristique durable*. Éditions Universitaires Européennes, Chisinau.

Afeissa, H.S. (2007). *Éthique de l'environnement. Nature, valeur, respect.* Vrin, Paris.

Barthes, A. and Alpe, Y. (2012). Les éducations à, un changement de logique éducative ? L'exemple de l'éducation au développement durable à l'université. *Spirale. Revue de recherches en éducation*, 50, 197–209.

BITS (1972). Vienna Charter. BITS, Brussels.

Breton, J.M. (ed.) (2001). *L'écotourisme, un nouveau défi pour la Caraïbe ?* Khartala, Paris.

Buckley, R. (2008). *Ecotourism. Principles and Practices*. CABI, London.

Buckley, R. (ed.) (2010). *Conservation Tourism*. CABI, London.

Buckley, R., Pickering, C., Weaver, D.B. (eds) (2008). *Nature-Based Tourism, Environment and Land*. CABI, London.

Butler, R.W. (1980). The concept of a tourist area cycle of evolution: Implications for management of resources. *The Canadian Geographer*, 24(1), 5–12.

Caccomo, J.L. (2007). *Impact des activités touristiques sur l'environnement. Fondements d'économie du tourisme. Acteurs, marchés, stratégies*. De Boeck Supérieur, Brussels.

Cazes, G. (1992). *Fondements pour une géographie du tourisme et des loisirs*. Bréal, Rosny.

Coke-Hamilton, P. (2020). Le coronavirus coûtera au tourisme mondial au moins 1.200 milliards de dollars (CNUCED) [Online]. Available at: https://news.un.org/fr/story/2020/07/1072312 [Accessed 3 August 2020].

Diop, C.A. (1981). *Civilisation ou Barberie*. Présence africaine, Paris.

Dumazedier, J. (1962). *Vers une civilisation du loisir ?* Le Seuil, Paris.

Faouzi, H. and Martin, J. (2014). Soutenabilité de l'arganeraie marocaine. *Confins*, 20.

François, H. (2004). Le tourisme durable : une organisation du tourisme en milieu rural. *Revue d'économie régionale et urbaine*, 1, 57–80.

Garnier, B. (2014). Territoires, identités et politiques d'éducation. *Carrefours de l'éducation*, 38, 127–157.

Garnier, B. (2018). L'éducation informelle contre la forme scolaire. Note de synthèse. *Carrefours de l'éducation*, 45, 67–91.

Gauthier, B. and Bourgeois, I. (eds) (2016). *Recherche sociale. De la problématique à la collecte des données*. PUQ, Quebec.

Hall, N. (2003). Écotourisme, tourisme durable, tourisme responsable ou tourisme équitable ? L'Ère de l'écotourisme. Special bulletin, Association québécoise pour la promotion de l'éducation relative à l'environnement (AQPERE).

Laliberté, M. (2005). Le tourisme durable, équitable, solidaire, responsable, social … : un brin de compréhension. *Téoros*, 24(2), 69–71.

Lequin, M. (2001). *Écotourisme et gouvernance participative*. PUQ, Quebec.

Lequin, M. (2002). L'écotourisme. Expérience d'une interaction nature-culture. *Téoros*, 38, 38–42.

Munro, D.A. (1991). *Caring for the Earth: A Strategy for Sustainable Living*. UICN-PNUE, Gland.

OMT (1999). Code mondial d'éthique du tourisme. General Assembly – Thirteenth session, Santiago, September 27–October 1, resolution A/RES/406(XIII).

Orams, M.B. (1995). Towards a more desirable form of ecotourism. *Tourism Management*, 16(1), 3–8.

Preuil, S. (2015). Le développement touristique des “4 000 îles” au Laos. Quels compromis entre développement économique et conservation dans les petites îles fluviales fragiles. In *Espaces et environnements littoraux et insulaires. Accessibilité, vulnérabilité, résilience*, Breton, J.M., Dehoorne, O., Furt, J.M. (eds). Karthala, Paris.

Rogers, A. (2004). *Non-Formal Education. Flexible Schooling or Participatory Education? University of Hong Kong, Comparative Education Research Center*. Kluwer Academic Publishers, Hong Kong.

Sarrasin, B., Dehoorne, O., Augier, D. (2016). Tourisme et ressources naturelles, prendre la mesure du défi. *Études caribéennes*, 33–34.

Shackelford, P. (2002). Sustainable tourism in the post Rio context, principles and programmes [Online]. Available at: http://www.insula.org/tourism/omt2.htm [Accessed 3 August 2020].

Tarrier, M. and Benzyane, M. (2003). L’arganeraie marocaine se meurt : problématique et bio-indication. *Sécheresse*, 1(1) [Online]. Available at: http://www.secheresse.info/article.php3?id_article=228 [Accessed 3 August 2020].

UNESCO and UNEP (1987). Projet de rapport final de la conférence de Moscou. Report, UNESCO and UNEP.

Vincent, G. (1980). *L’école primaire française. Étude sociologique*. Presses universitaires de Lyon, Lyon.

Zaoual, H. (2006). Développement, organisations et territoire : une approche Sud-Nord. *Innovations*, 24(2), 9–40.

10

Biosphere Reserves and Political Skills Transfer in University Curricula

10.1. Introduction

While scientific warnings are old (Meadows report (1972), etc.), those concerning the rapid erosion of biodiversity have been taking shape for two decades, to the point of describing the current situation as the sixth major mass extinction crisis (Thomas et al. 2004), the primary cause of which is the conflict over the use of spaces and resource overexploitation ("Living Planet" report, WWF). Since then, various policies described as fostering "sustainable development" have supported mechanisms formalizing the links between the necessary management of biodiversity (Blandin 2009; Rockström et al. 2009) and territorial development. They are relayed by globalized educational policies (UNESCO's Decade of Education for Sustainable Development (2004–2014); the Education Roadmap 2030, UNESCO 2015–2030), ultimately joining higher education curricula.

In recent mechanisms, the importance of integrated scientific approaches to globalized social and political issues is systematically recalled. Higher education, regularly called upon to take charge of the major transverse issues of society, follows in the footsteps of the mechanisms implemented.

Since 1971, the UNESCO Man and the Biosphere (MAB) Programme has fostered different articulation principles for the political, scientific and academic spheres in various territories. Theoretically, the declared goal is to reconcile sustainable development at the regional level with the protection of the environment and, more specifically, the conservation of biodiversity with respect for the cultural values of

---

Chapter written by Melki SLIMANI, Angela BARTHES and Jean-Marc LANGE.

all. Biosphere reserves, which are the formal UNESCO label, are considered as privileged territorial spaces for experimenting with specific operating methods in transverse goals and issues, as well as for accompanying them with appropriate educational systems. The different global and local political scales are articulated, with various consequences on the reconfiguration of local political arenas (Barthes and Tebbaa 2019); on specific modes of development regarding the renewal of bonds between knowledge, powers and institutions; on renovated relationships between the scientific and educational spheres; and on the governance of territories.

We postulate that university degrees supporting international mechanisms of this type also vivify international policies. In addition to disciplinary teaching, there are other types of transverse teachings with a reflective and political nature. In these diplomas, part of the teaching also relates to education in politics. Even though this is not necessarily mentioned as such, it fulfills the ideological function of a hidden curriculum (Barthes and Alpe 2018). While this teaching is derived from an international political order, it also responds to a social demand (Lange 2011a, 2011b), leading to curricular hybridization (Barthes and Alpe 2018) between the informal, the non-formal (Gasse 2017) and the formal.

The development of this type of hybrid content has been in full development since the beginning of the 2000s in new and diverse fields such as health, risk prevention, citizenship, environment, sustainable development, gender equality, etc. It takes on the major challenges of society and aims to create the conditions for the acquisition of skills that will enable young people to become social and political actors (Garnier 2018). Thus, the construction of hybrid curricula mainly encourages the development of a recontextualized political skill in a world on the verge of an impending ecological crisis. Such skill makes use of contributing knowledge (Lebeaume 2019) in the environmental field acquired in the social sphere (drawn from disciplinary fields such as environmental humanities or environmental social sciences) .

We propose the epistemological formalization of the political skill in university curricula. We focus on the question of contributing knowledge, encouraging the development of this skill. Our analysis is based on the case study of a master's degree inspired by UNESCO's MAB Programme.

10.2. Towards a conceptual recontextualization of the political skill in the environmental field

The political skill is defined as the ability or disposition to perform a political action. It differs from political prudence in that it is a purely instrumental skill, in other words, the ability to get what everyone wants in politics. In point of fact, the political skill also refers to the ability or disposition to identify the grave injustices that pervade human societies (Brinkmann 2018).

To develop a conceptualization of the political skill that can be approached by university curricula in the form of environmental and development issues (EDIs), we focus on the conceptual political skill component framework developed by Sloam (2010, 2011) in the context of the academic teaching of political science. The author identifies this skill with four types of mutually reinforcing learning processes: political literacy, political participation, political deliberation and political experience learning. By transposing it to the environmental domain, we consider that the environmental political skill involves learning eco-literacy, eco-citizenship, environmental deliberation and experiential collective action.

We also relied on a comprehensive synthesis of the educational purposes of socialization (Slimani et al. 2018) to establish the conceptual links between these learning processes and the construction of a disposition towards socialization. These provisions correspond to a set of proactive attitudes, the appropriation of significant concepts and the capabilities for taking action, gradually regulated by prospective (and sometimes explanatory) knowledge. They combine attitudes and habits, abilities and skills, in particular the attitudes and capacities for co-operation, engagement, audience judgment and the evaluation of negative externalities resulting from public action (Lange 2011a, 2011b, 2014). They define what Bourdieu (1998) calls a *habitus*, which is more or less consciously acquired by an agent, a member of a socially constituted group, and can act as a matrix orienting individual practices and ways of thinking.

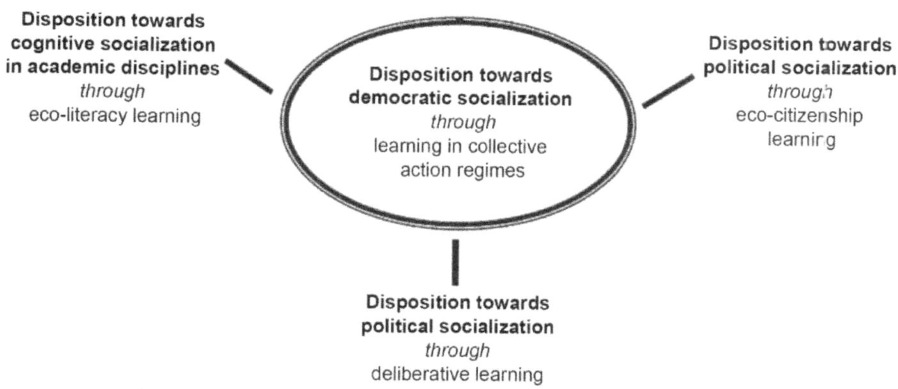

Figure 10.1. *Conceptual framework of the environmental political skill*

This allows us to conceptually differentiate the political skill in educational contents from the disposition towards socialization. Thus, cognitive socialization in academic disciplines is related to eco-literacy learning, political socialization to eco-citizenship learning, critical cognitive socialization to deliberative learning and democratic socialization to learning in collective action regimes (Figure 10.1).

10.2.1. *Learning eco-literacy and building the disposition towards cognitive socialization in academic disciplines*

Eco-literacy learning refers to learning scientific disciplines that involve environmental literacy as an element of scientific literacy. Huckle (2013) identified three complementary forms of this type of literacy:

– technical eco-literacy evokes an understanding of the fundamental sciences, assessing their relevance to societal issues;

– cultural eco-literacy involves understanding the ways in which different cultures interpret and value nature;

– critical eco-literacy evokes an understanding of sustainable development in a context of political economy and of social production and reproduction.

Thanks to the theory of didactic situations, learning in academic disciplines can be interpreted as a cognitive adaptation to different learning environments (Orange 2007). Such learning environments constitute acculturation situations (Albe 2011) in scientific practices, which in turn lead to the socialization of learners through disciplinary matrices and their cognitive logics (Millet 2013).

10.2.2. *Learning eco-citizenship and building the disposition towards political socialization*

Learning citizenship involves learning social roles in a community of eco-citizens emerging from a requirement to reduce the ecological footprint (Dobson 2006; Machin 2012) in its various components (water, energy, carbon, etc.). Eco-citizenship can be identified with a political sphere resulting from the relationship of the human community to the environment where different types of learning are developed, such as knowing how to act politically (knowing how to denounce, to resist, to choose, to claim, to propose, to commit oneself, to create projects) and how to act collectively (Sauvé 2017).

This type of learning thus establishes political socialization. Synthesizing conceptions on the "ideal citizen" from a critical pedagogical perspective, Johnson and Morris (2010) identified three conceptions of political citizenship: adaptive, individualistic and critical-democratic. The adaptive citizen is the one who is obedient, has good manners and acts responsibly. The individualistic citizen participates in society by adopting an individualistic perspective. The critical democratic citizen is co-operative, concerned with social justice and motivated for social change. We consider eco-citizenship as a form of political citizenship; hence, we propose using the same analytical framework as Johnson and Morris.

10.2.3. *Learning environmental deliberation and building the disposition towards critical cognitive socialization*

Learning political deliberation is learning how to debate in problematic environmental situations (Machin 2014; Håkansson et al. 2018) plagued by scientific uncertainty. Håkansson et al. (2018) differentiated these situations, in the context of environmental and sustainable development education, depending on the "political tendency" underlying them and fostering socialization in democratic values and attitudes. The authors distinguished three categories of political deliberation: normative (oriented towards the adoption of a point of view a priori specified by the teacher, following a dominant orientation), consensual (oriented towards consensus to overcome divergences) and conflictual (taking up the defense of opposing and challenging perspectives).

Debate is a tool that promotes argumentation and develops an exploratory discourse, opening up scientific controversies to the fields of possibilities (Morin 2013). This tool becomes a critical means of socialization (Simonneaux and Legardez 2008) because it guarantees a certain distance from human reality.

10.2.4. *Learning in collective action regimes and building the disposition towards democratic socialization*

Learning in collective action regimes is a political experience based on collective environmental actions carried out at the level of the institution and its territory. Taking into account the work on the sociology of organizations, Lange (2011a, 2011b, 2015) identified three types of collective action regimes. The first is a minimalist regime, in which the co-ordination between social agents is only a juxtaposition of interventions (inter-comprehension is supposed to be instantaneous). The second is a maximalist regime, in which co-ordination forms a learning community (sharing of judgments relating to goals). In an intermediate regime, co-ordination can take the form of a community of practice.

The author associates these action regimes with a reflexivity relationship the actors maintain towards their action and that of others in the form of collective action. Such reflexivity participates in the construction of "a disposition" resulting from a socialization mode based on discussion and on the indeterminateness of the future: this is democratic socialization (Vincent 2008).

We conceptualize the political skill in the environmental educational field to describe the potentialities of the contents of a master's degree inspired by the UNESCO MAB Programme for developing such a skill. Through curricular didactics, we examine the contributions from environmental humanities and environmental social sciences to this skill.

10.3. Environmental political skill: Master's degree in Man and the Biosphere – case study

10.3.1. *The MAB Programme, political skill and formal curriculum*

Education is a major goal of UNESCO's MAB Programme, which has a focus on conservation and research. It is thanks to this educational purpose that biosphere reserves, deployed in the context of this program, constitute spaces for raising awareness and for local participation in a management policy. These spaces are truly spaces for political education thanks to the conflict resolution mechanisms integrated into their geographical structuring. They contain spaces for co-operation with the population, with a soft stroke, in addition to conservation and scientific research areas (Cibien 2006). Empirical research has shown the educational impact of these sites not only on the political participation of local populations (Ishwaran et al. 2008; Schultz and Lundholm 2010) but also the reorganization modes of local political arenas in UNESCO systems (Barthes and Tebbaa 2019). It thus appears that the development of a political skill among these populations constitutes a strategic orientation in UNESCO's MAB Programme.

The educational aspect of the MAB Programme has gained access to formal university education in the form of a master's degree course. This has provoked certain tensions in the formal curriculum, which is often dependent on academically recognized disciplines and social demand. This hybridization (Barthes and Alpe 2018) leads to the curricular construction of the political skill, now made possible via the formal curriculum form which selects, prioritizes and legitimizes content (Alpe and Barthes 2013; Barthes 2017). In this co-construction, knowledge is designed and engaged in a functional way, inviting contributions from various academic disciplines. These contributions can take the form of complex disciplinary and interdisciplinary reconfigurations (Lange 2011b, 2017a, 2017b). At the same time, this poses epistemological, pedagogical and didactic problems, as well as risks of relativist and absolutist drifts (Barthes and Alpe 2018).

The development of a formal curriculum oriented by the MAB Programme fundamentally appears to be a political issue.

The descriptions formalized in this way make it possible to propose methods for analyzing political skills in higher education curricula.

10.3.2. *Methodology*

10.3.2.1. *First stage: Construction of an analysis grid for the political skill in the curricula*

We exploit the conceptualization of the political skill in the field of environmental education so as to construct an analysis grid (Table 10.1). It describes the potential of educational contents for developing that skill (Slimani et al. 2017; Slimani 2019, 2020). Then, the contents of the diplomas are read and the elements of vocabularies are noted to serve as indicators of the presence or absence of the predefined analysis criteria. These indicators constitute the components of the political skill of this training program.

Components of the environmental political skill		**Indicators**
Disposition towards cognitive socialization in academic disciplines	Eco-literacy learning	– Technical: fundamental sciences and their relevance to societal issues – Cultural: valorization of nature depending on the culture – Critical: political economy, social reproduction
Disposition towards political socialization	Learning eco-citizenship	– Adaptive: good manners, obedience, responsible actions – Individualistic: social participation with an individualist perspective – Critical-democratic: co-operation, social justice, motivation for social change
Disposition towards critical cognitive socialization	Learning deliberation	– Normative: a priori specified points of view – Consensual: consensus overcoming differences – Conflictual: defense of opposing perspectives
Disposition towards democratic socialization	Learning in collective action regimes	– Minimalist: separate interventions by social actors – Intermediate: community of practice – Maximalist: learning community

Table 10.1. *Indicators of the environmental political skill in educational contents involving EDIs*

This approach will help us define an ideal type for the potential for acquiring the political skills in the training program, depending on the absence or presence of the components observed (Figure 10.2).

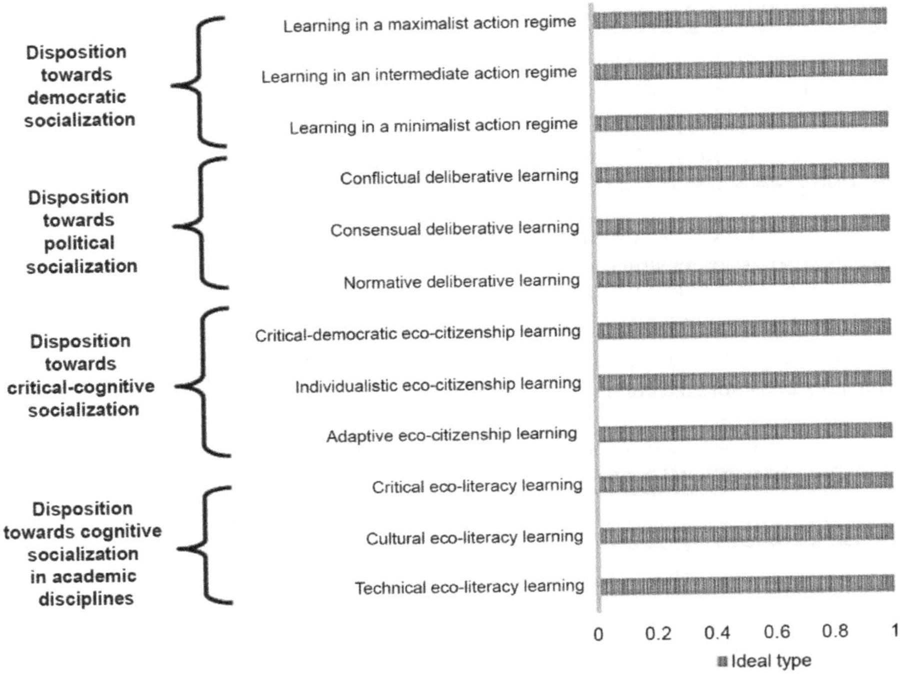

Figure 10.2. *"Disposition towards socialization" ideal types generated by political learning through EDIs*

10.3.2.2. *Second stage: Analysis of curricular morphologies, contributing knowledge and constitutive practices*

Scientific knowledge relating to environmental issues in the fields of the environmental humanities and the environmental social sciences is a key resource for developing an environmental political skill. We propose completing the analysis grid of the environmental political skill with the analysis of curricular morphology (Lebeaume 2019).

It will suffice to seek the contributions from environmental humanities and the environmental social sciences in the development of the different components. For this, we refer to Lebeaume's (2019) idea that functionally differentiates constitutive knowledge from contributing knowledge when describing the structure of the curricula or repositories for professionalization in transverse areas. Constituent practices functionally fall under the *objective-driven* or *process-driven* register, whereas contributing knowledge functionally belongs to the *content-driven* register.

From the analytical framework posed by Lebeaume, we cross the provisions of the skills policies with contributing disciplines and constitutive practices in a table. This defines the curricular morphology of diplomas and can be used for its graphical representation.

10.3.2.3. *Case study*

The contents chosen for the analysis of the political skill were drawn from the formal curriculum of the master's degree at Paul Sabatier University in Toulouse (Toulouse III). This formal curriculum was compiled as a syllabus for the 2019/2020 academic year under a master's accreditation period which extended from 2016 to 2021.

This training program is the only one in the Mediterranean area explicitly in line with the spirit of the UNESCO MAB Programme. The syllabus of the master training in biodiversity, ecology and evolution is divided into two parts. The first part, comprising about 60 pages, includes the contents of the training program during the first year or common core (M1 syllabus). The second part (about 20 pages) refers to the training contents during the second year (M2 syllabus). It explicitly describes the course as an MAB training program. Focusing on the syllabus, we first look for political skill potentialities and then we search for the contributions from constituent disciplines and practices, the whole of which gives shape to curricular morphology.

10.4. Results and discussion

10.4.1. *What political skill in the formal curriculum of the MAB master's degree?*

Analysis of the contents assigned during the two years of training enabled us to collect indicators of the possible political lessons in this curriculum form and subsequently to characterize the provisions of political skill conceivable in its content (Table 10.2). Figure 10.3 summarizes the components of the environmental political skill in this curriculum.

Components of the environmental political skill		**Indicators collected from the formal curriculum of the MAB master's degree**
Disposition towards cognitive socialization in academic disciplines	Eco-literacy learning	Technical: – the physico-chemical functioning of aquatic systems, the impact of human activities on the structure and functioning of aquatic systems (M1 syllabus, p. 15) – understanding formation mechanisms, the functioning and dynamics of (plant and animal) species assemblages in terrestrial communities … (M1 syllabus, p. 16) – the goal is to acquire a solid knowledge base on diversity, structure, functioning, geological history, ecology and the evolution of tropical ecosystems, as well as human–forest interactions in the tropics (M1 syllabus, p. 31) – being able to use environmental data while exploring territorial themes related to the environment, biodiversity management and/or land planning … (M2 syllabus, p. 12)
		Critical: at the end of this course, students should be able to account for the social, technical and political issues regarding the evolution of relations between societies and the environment
Disposition towards political socialization	Learning citizenship	Adaptive: awareness while assessing the natural and socio-economic patrimonial value of a site and development of practices for the management of species or natural spaces (M2 syllabus, p. 17)
		Individualistic: – protected areas and biosphere reserves: thinking about territories to reconcile the conservation of biodiversity – human well-being (M2 syllabus, p. 13)
Disposition towards political socialization	Learning citizenship	Critical-democratic: – social and solidarity economy for the conservation of biodiversity (M2 syllabus, p. 13) – taking ownership of the issues, practices and limitations of participatory approaches aimed at engaging different types of actors for concerted management of territories (M2 syllabus, p. 14)
Disposition towards critical cognitive socialization	Learning deliberation	Normative: […] conveying a message of a scientific and technical nature to different types of audiences (naturalist associations, design offices, administrations, communities, general public) (M2 syllabus, p. 19)
Disposition towards democratic socialization	Learning in collective action regimes	Minimalist: […] presentation of analytical frameworks to understand the relationships between actors with diverse interests, knowledge and practical experience within a territory
		Maximalist: – construction of sustainable development education projects to engage and involve a variety of audiences (M2 syllabus, p. 11) – co-construction of a representation of a complex socio-ecological system, hosting sessions to reach concerted action, project development (M2 syllabus, p. 17)

Table 10.2. *Indicators of the environmental political skill in the formal curriculum of the MAB master's degree at Paul Sabatier University in Toulouse*

10.4.2. *What disciplinary contributions to the environmental political skill in the formal curriculum of the MAB master's degree?*

The intersection between these two registers in the case of the contents of the formal curriculum of the MAB master's degree helps us to outline the morphology of this curriculum (Figure 10.4).

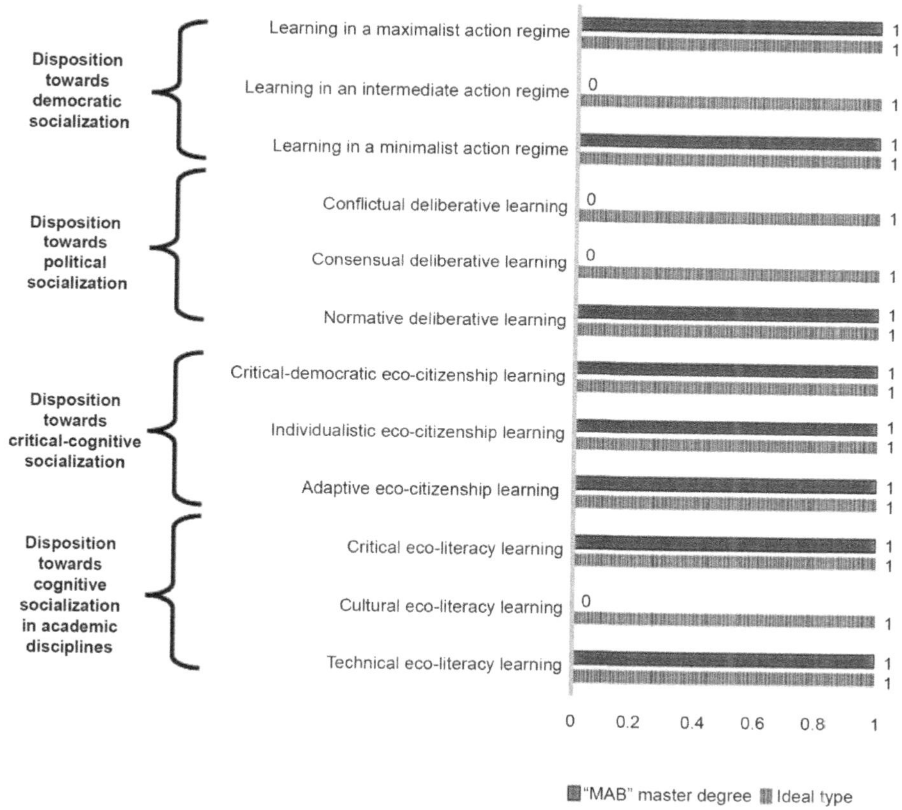

Figure 10.3. *The environmental political skill in the formal curriculum of the MAB master's degree*

The morphology of the formal curriculum of the MAB master's degree shows that, in addition to the life sciences, five fields of knowledge in the social sciences appear as essential contributors for building up the environmental political skill. It also associates the specific approaches integrating conflict management and role-playing.

The first field is that of the sociology of the environment, taking the form of a teaching unit with the same title and contributing to the construction of a disposition towards democratic socialization, incorporating learning in a maximalist action regime.

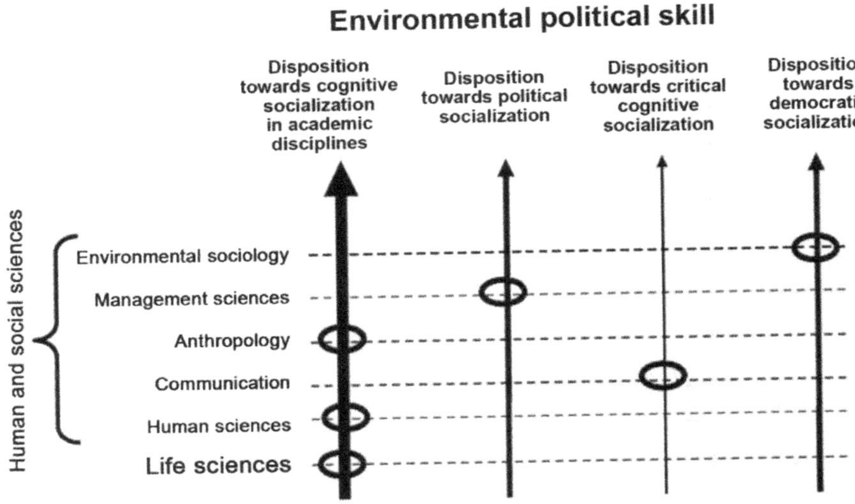

Figure 10.4. *Contributions from the social sciences to the environmental political skill within the MAB master's degree*

The second field is that of management sciences under a teaching unit called "Protected Areas and Biosphere Reserves". This contributes to the construction of a disposition towards political socialization by learning critical citizenship. This learning is committed to social justice because it blends social and solidarity economy with the idea of ecological solidarity.

The third field is that of anthropology (as related to biology and ecology) under a teaching unit called "Anthropobiology and Human Ecology". The contribution of this field focuses on technical eco-literacy learning.

The fourth field is that of communication under a teaching unit called "Communication Practice". This contributes to the construction of a disposition towards critical cognitive socialization thanks to deliberative learning. However, this does not go beyond the level of normative deliberation.

The fifth field is that of the humanities associated with the social sciences under a teaching unit called "Human and Social Sciences Perspectives on Science–

Society–Environment Interactions", which focuses on learning critical eco-literacy. We note the absence of possible contributions from the field of the humanities[1] (literature, art and philosophy).

This morphology also shows a focus on dispositions towards cognitive socialization in the academic disciplines of life sciences through eco-literacy learning. This study has shown the importance of the contributions from interdisciplinary fields of knowledge such as environmental social sciences. These are the essential contributors for the other components of the environmental political skill (political socialization, democratic socialization and critical cognitive socialization). It seems that the absence of contributions from the environmental humanities (literary eco-criticism, environmental philosophy and environmental art), valuing and appreciating diversity and differences, has limited the possibilities of aiming for the highest disposition levels towards critical cognitive socialization via the learning of conflictual deliberation (between non-enemy adversaries).

In the formal curriculum of the MAB master's degree, the contents relating to the construction of dispositions towards cognitive socialization in academic disciplines are preponderant. It gathers the contributions from the disciplinary field of life sciences, with a lesser contribution from the human sciences and anthropology. The other three types of disposition involve contributions from the fields of sociology, the environment, communication and management sciences.

10.5. Conclusion: changing curricular morphologies

As part of the management of major societal challenges by higher education, curricula are hybrids (Barthes 2017) and their morphologies evolve, on the one hand, through the skill-based approach, and on the other hand, through the more assertive involvement of the political dimension. It targets the development of a disposition cultural capital (Morlaix 2015), or more generally, a habitus capable of generating action capabilities in this new context. University curricula, reconfigured by international prescriptions (Barthes and Alpe 2018), are shifting towards the construction of contents developing disciplinary and non-cognitive (transverse)

---

1 We draw a distinction between social sciences and the humanities. The former (sociology, political science, economics, etc.) are oriented towards the search for regularities by approaching the natural sciences through the adapted use of their methodologies. The humanities (philosophy, literature, art, history, etc.) value and appreciate creativity and differences. We also distinguish between human sciences (psychology, anthropology, etc.), which are methodologically much closer to the social sciences (forming the social sciences with them), and the humanities.

cognitive capacities, such as the disposition towards questioning environmental injustices (Levinson et al. 2020).

The environmental political skill at the Anthropocene age clearly articulates the dispositions that can be envisaged in educational (learning) processes of socialization, fighting against injustices with the help of a transverse dimension: the critical dimension. This dimension can be worked by learning critical eco-literacy, critical democratic eco-citizenship and conflictual deliberation, all of which can be articulated by learning in a regime of maximalist collective action.

UNESCO's MAB Programme provides a singular tool, the biosphere reserves, offering an explicit educational dimension for socialization. This program and its mechanisms can be considered as a reference social practice (Martinand 2018) for the development of the formal academic curriculum of the MAB master's degree. This is an important factor favoring the entry of political skills in the organization of curricular contents.

10.6. References

Albe, V. (2011). Finalités socio-éducatives de la culture scientifique. *Revue française de pédagogie*, 174, 119–138.

Alpe, Y. and Barthes, A. (2013). De la question socialement vive à l'objet d'enseignement : comment légitimer des savoirs incertains ? Les sciences et les crises contemporaines. *Les dossiers de l'éducation*, 29, 33–44.

Barthes, A. (2017). Quels outils curriculaires pour des "éducations à" vers une citoyenneté politique ? *Éducations*, 17(1), 25–40 [Online]. Available at: https://www.openscience.fr/Numero-1-327.

Barthes, A. and Alpe, Y. (2018). Les "éducations à", une remise en cause de la forme scolaire. *Carrefours de l'éducation*, 45, 13–21, 23–37.

Barthes, A. and Tebbaa, O. (2019). Savoirs et conflits de savoirs en éducation à l'environnement et au développement durable : le cas des dispositifs éco-orientés UNESCO. *Éducation relative à l'environnement*, 15(1) [Online]. Available at: http://journals.openedition.org/ere/3657.

Barthes, A., Lange, J.M., Tutiaux-Guillon, N. (2017). *Dictionnaire critique des enjeux et concepts des "éducations à"*. L'Harmattan, Paris.

Bourdieu, E. (1998). *Savoir faire : contribution à une théorie dispositionnelle de l'action.* Le Seuil, Paris.

Blandin, P. (2009). *De la protection de la nature au pilotage de la biodiversité.* Quae, Versailles.

Brinkmann, M. (2018). The concept of political competence. *Critical Review*, 30, 163–193.

Cibien, C. (2006). Les réserves de biosphère : des lieux de collaboration entre chercheurs et gestionnaires en faveur de la biodiversité. *Nature Sciences Société*, 14, 84–90.

Dobson, A. (2006). Ecological citizenship: A defence. *Environmental Politics*, 15(3), 447–451.

Garnier, B. (2018). L'éducation informelle contre la forme scolaire. *Carrefours de l'éducation*, 45, 13–21.

Gasse, S. (2017). Éducation informelle. In *Dictionnaire critique des enjeux et concepts des "éducations à"*, Barthes, A., Lange, J.M., Tituaux-Guillon, N. (eds). L'Harmattan, Paris.

Håkansson, M., Östman, L., Van Poeck, K. (2018). The political tendency in environmental and sustainability education. *European Educational Research Journal*, 17(1), 91–111.

Huckle, J. (2013). Eco-schooling and sustainability citizenship: Exploring issues raised by corporate sponsorship. *The Curriculum Journal*, 24(2), 206–223.

Ishwaran, N., Persic, A., Tri, N.H. (2008). Concept and practice: The case of UNESCO biosphere reserves. *International Journal of Environment and Sustainable Development*, 7(2), 118–131.

Johnson, L. and Morris, P. (2010). Toward a framework for critical citizenship education. *The Curriculum Journal*, 12(1), 77–96.

Lange, J.M. (2011a). Éducation au développement durable : problématique éducative/problèmes de didactique. Thesis, École normale supérieure, Cachan.

Lange, J.M. (2011b). Penser en termes de contribution l'éducation scientifique à l'éducation au développement durable. *Revue des Hautes écoles pédagogiques*, 13, 137–156.

Lange, J.M. (2014). Des dispositions des personnes aux compétences favorables à un développement durable : place et rôle de l'éducation. In *Éducation au développement durable, enjeux et controverses*, Diemer, A. and Marquat, C. (eds). De Boeck Supérieur, Brussels.

Lange, J.M. (2015). Éducation et engagement. La participation de l'école à relever les défis environnementaux et de développement. *Éducation relative à l'environnement : Regards, Recherches, Réflexions*, 12, 105–125.

Lange, J.M. (2017a). Éducation au développement durable : enjeux épistémologiques et didactiques des reconfigurations disciplinaires possibles. In *Curricula et disciplines scolaires : changements et reconfigurations*, Vergnolle-Mainar, C. (ed.). Presses Universitaires du Midi, Toulouse.

Lange, J.M. (2017b). Curriculum. In *Dictionnaire critique des enjeux et concepts des "éducations à"*, Barthes, A., Lange, J.M., Tutiaux-Guillon, N. (eds). L'Harmattan, Paris.

Lange, J.M. and Martinand, J.L. (2010). Éducation au développement durable et éducation scientifique : repères pour un curriculum. In *Enjeux contemporains de l'éducation scientifique et technique*, Hasni, A. and Lebeaume, J. (eds). Les Presses de l'Université d'Ottawa, Ottawa.

Lebeaume, L. (2019). Précisions sur la "forme curriculaire" et distinction entre pratiques constitutives et savoirs contributifs. *Éducation et didactique*, 13(1), 43–59.

Levinson, R., Paraskeva-Hadjichambi, D., Bedsted, B., Manov, B., Hadjichambis, A.C. (2020). Political dimensions of environmental citizenship. In *Conceptualizing Environmental Citizenship for 21st Century Education. Environmental Discourses in Science Education*, Hadjichambis, A., Reis, P., Paraskeva-Hadjichambi, D., Činčera, J., Boeve-de Pauw, J., Gericke, N., Knippels, M.-C. (ed.). Springer, Berlin.

Machin, A. (2012). Decisions, disagreement and responsibility: Toward an agonistic green citizenship. *Environmental Politics*, 21(6), 847–863.

Machin, A. (2014). Ends, means, beginnings: Environmental technocracy, ecological deliberation or embodied disagreement? *Ethical Perspectives*, 21(1), 47–72.

Martinand, J.L. (2018). Pratiques sociales de références et autres concepts. *Économie et Management*, 168, 68–73.

Millet, M. (2013). Les cadres cognitifs de la socialisation. Thesis, University of Poitiers, Poitiers.

Morin, O. (2013). Éducation à la citoyenneté et construction collaborative de raisonnements socio-scientifiques dans la perspective de la durabilité : pédagogie numérique pour une approche interculturelle de QSV environnementales. University of Toulouse-Jean Jaurès, Toulouse.

Morlaix, S. (2015). Les compétences sociales à l'école primaire : essai de mesure et effets sur la réussite. *Carrefours de l'éducation*, 40, 183–200.

Orange, C. (2007). Quel milieu pour l'apprentissage par problématisation en sciences de la vie et de la Terre. *Éducation et Didactique*, 1(2), 37–55.

Rockström, J., Steffen, W., Noone, K., Persson, Å., Chapin III, F.S., Lambin, E.F., Lenton, T.M., Scheffer, M., Folke, C., Schellnhuber, H.J. et al. (2009). A safe operating space for humanity. *Nature*, 461, 472–475.

Sauvé, L. (2017). Éducation à l'écocitoyenneté. In *Dictionnaire critique des enjeux et concepts des "éducations à"*, Barthes, A., Lange, J.M., Tutiaux-Guillon, N. (eds). L'Harmattan, Paris.

Schultz, L. and Lundholm, C. (2010). Learning for resilience? Exploring learning opportunities in biosphere reserves. *Environmental Education Research*, 16, 645–663.

Simonneaux, L. and Legardez, A. (2008). Efficacité de l'approche des Questions Socialement Vives pour l'éducation à l'environnement et à la durabilité. In *Symposium au colloque efficacité et équité en éducation* [Online]. Available at: https://esup.espe-bretagne.fr/efficacite_et_equite_en_education/programme/symposium_legardez_lsimonneaux.pdf.

Simonneaux, L., Panissal, N., Brossais, E. (2013). Students' perception of risk about nanotechnology after an SAQ teaching strategy. *International Journal of Science Education*, 35, 2376–2406.

Slimani, M. (2019). Vers un curriculum possible d'une éducation au politique au travers des questions d'environnement et de développement (QED). PhD thesis, University of Montpellier and Virtual University of Tunis, Montpellier and Tunis.

Slimani, M. (2020). Balises curriculaires pour une éducation au politique à travers les questions environnementales et de développement (QED). Communication, *Deuxièmes journées d'étude "curriculum"*, University of Montpellier, Montpellier.

Slimani, M., Lange, J.M., Azzouna, A. (2017). Questions environnementales et de développement et transitions vers la durabilité : quels contenus éducatifs pour le développement d'une littératie politique ? *Actes du Colloque international "Changements et Transitions : enjeux pour les éducations à l'environnement et au développement durable"*, Toulouse, 183–194.

Slimani, M., Lange, J.M., Azzouna, A. (2018). Le projet pédagogico-didactique de socialisation : perspective d'une socialisation démocratique pour l'enseignement-apprentissage des questions environnementales et de développement. *Éducation et socialisation*, 50 [Online]. Available at: https://journals.openedition.org/edso/5106.

Sloam, J. (2010). Youth, citizenship, and political science education: Questions for the discipline. *Journal of Political Science Education*, 6, 325–335.

Sloam, J. (2011). A pedagogy of engagement in higher education. *Meeting annuel de l'Association Américaine de Science Politique*, 1–4 September, Seattle.

Thomas, C.D., Cameron, A., Green, R.E., Bakkenes, M. (2004). Extinction risk from climate change. *Nature*, 427, 145–148.

Paul Sabatier University (2019). Syllabus Master : Mention biodiversité, écologie et évolution, M1 biodiversité, écologie, évolution.

Vincent, G. (2008). La socialisation démocratique contre la forme scolaire. *Éducation et francophonie*, 36(2), 47–62.

11

Education and Mediation in the Arganeraie: Alliance Strategies Between Education and Tourism Actors?

11.1. Introduction

Territorial issues challenge scientists and everyone interested in the UNESCO World Heritage, in particular; the articulations between biosphere reserves and the Sustainable Development Goals 2030. These involve delicate choices between several alternatives including territorial, human and sustainable impacts, as well as options resulting from deliberate policies, in order to meet new global challenges.

In Morocco, as in other world regions, a share of the responsibility for development concerning the UNESCO World Heritage falls on local, regional, national and international authorities. This responsibility sharing in the new decentralization and deconcentration policy can reveal a lack of skill or capacity on the part of patrimony-based education actors.

How is the Arganeraie biosphere reserve (ABR) mediated in the virtual world in the absence of any school-specific learning in manuals or in educational notebooks? What media should be used (films, photos, texts, specific sites, science, advertising, etc.)? What content to communicate? Who manages this type of communication and by what means? What is the role of training for improving and developing appropriate communication?

Chapter written by Saïd BOUJROUF and Abdullah AÏT L'HOUSSAIN.

11.2. Locating the Arganeraie biosphere reserve

Biosphere reserves (BRs) are regions recognized by UNESCO in its goal to strengthen the local capacity for the protection of cultural and natural patrimony. In fact, biosphere reserves seal a pact between the community and society as a whole, which can be interpreted as an informal social contract.

The ABR was created in 1998, spreading over 8,000 square kilometers around Morocco's endemic forest species: the argan tree (*Argania spinosa*). It is found in the Mediterranean–Saharan transition zone across a varied topography. It covers both plains and plateaus (Souss, Massa and Tiznit), as well as the surrounding mountains (the High Atlas in the north), reaching a maximum altitude of 2,000 m in the Anti-Atlas in the south.

To the west, the ABR borders with the Atlantic Ocean, forming the base of a pseudo-isosceles triangle that extends from the north of Essaouira 200 km south of Agadir and to the east of Taroudant. To the north, it is limited by the Tensift valley, to the south, by the Noun and Seyad valleys and to the east, it borders the surroundings of Aoulouz. The main physio-geographical units, as described in the framework plan of the ABR, are distributed as follows:

– northwest zone of the Atlantic High Atlas: plateaus of Chiadma-Haha and lower Tensift;

– the western High Atlas: from the Argana corridor to the coast and the southern slope of the High Atlas of Marrakech;

– the zone of the watershed of Souss, Massa and Tiznit;

– the western Anti-Atlas.

11.3. The ABR, a tourist landscape showcased by the media?

The question is not so much whether the ABR is perceived as patrimony but to understand the effects of media coverage for the valorization of biosphere reserves. Be they individuals or companies, the local inhabitants exploit images of the region with the goal of associating their products with a specific place (Protected Geographical Indication (PGI)).

The landscape is a showcase for attracting tourists, which gives a mark of quality to all the activities taking place within the ABR territory. It contributes added value to shops, crafts and local agricultural production, as well as the values associated with the rural area: ancestral know-how, the harmony between man and nature (incorporated into the product which came from nature).

The "terroir product" media operation launched by the Ministry of Agriculture as part of the Morocco's Green Plan Program in 2008 is proof of this. The Ministry granted official recognition to the rural areas, focusing on the impact of their economic development on the quality of agricultural products and landscapes. Are we witnessing a new mode of agricultural production which justifies the quality of its products not only by the anchoring in the soil, but also by the beauty of its landscape?

To a certain extent, the qualification of mediatized territories certifies the commitment of these actors to the protection of their landscape, the valorization and the maintenance of biodiversity. All these qualification procedures, such as, for example, the argan PGI, have the purpose of providing added value to the territory's actors (Romagny and Boujrouf 2010). They hope to promote the sustainable development of their natural patrimony. The commercial aspect of rural areas is mainly used by developers, local authorities and private bodies. Nevertheless, peasants are increasingly aware of the market value of the ABR. The arguments they use to justify intervention at the local level are in line with environmentalist and societal discourses (Levêque 2001). Still, in these labels, local actors find a way of bringing high added value to their productions. They can live with modest means in terms of cultivable land, herd size or degree of mechanization in their territory. We understand that a large number of producers have taken this path. The desire to preserve their traditional knowledge is less important compared to the economic interests represented by the ABR local products.

Just like the argan tree, the ABR landscape evokes the quality and resourcefulness of the people who shaped it. It reflects an ancestral know-how that modern agriculture has lost in several areas of the landscaped plain. The terraced landscape makes consumers understand that this is a rural area and authenticates the image of the terroir and traditional farming.

11.4. ABR landscape imaging and its dissemination

In addition to the four classic criteria that make up the ABR tourist landscape and the linguistic, literary (oral or written), pictorial and garden representations, does the appearance of photography introduce a fifth criterion?

Another revealer of the ABR tourist landscape today would be its imagery and its dissemination. It is under this form of mediation that it is imposed on society, widely disseminated by the Web, by postcards, leaflets from tourist offices and other documentation or digital devices. These stereotyped landscapes transform the identity of each region and contribute to forging aesthetic references.

Literary texts and photographic works play a big role in the invention of the ABR. By participating in the construction of the gaze, little by little they consecrate the ABR as an emblematic landscape. Holidaymakers also send their personal photographs to local inhabitants, which are used by the latter to illustrate their websites and reinforce mediation. This confers the places with a certain distance, creating a sort of landscape effect, which is not necessarily conveyed by the raw pictures. In a broader sense, was not the landscape merely "countryside" before people from elsewhere came to reveal it to the locals? We may wonder whether they have not surrendered to the seduction of novelty, delighted to imagine themselves in the role of smuggler, and justifying their insertion: for these newcomers, it is not a question of setting history in motion in order to inscribe their own history in it?

In the ABR, the "non-natives" are mainly new inhabitants who have chosen agriculture (the "neo-rurals"), the region's aesthetic characteristics being one of the first reasons for their presence, as if "the landscape's beauty" participated in their choice to come and live in a biosphere reserve.

Occasional residents or settled neo-rurals, seeking to imprint their presence in the place and get to know it, play the role of memory smugglers by carrying out real surveys of the locals, especially in the places they perceive as "patrimony", where human labor does not intervene so much (or rather where they do not perceive it). They then retransmit the place's history they have collected from the elders to the entire population.

The attribution of a purely aesthetic value to the living environment, its aestheticization, seems related to a distance regarding this space, to the possibility of taking an exogenous look at it. Does this mean that the people from these places cannot perceive their aesthetics or appreciate the beauty of the landscapes surrounding them? Or rather that their relationship to aesthetics is imbued with values, in particular the value of work on the landscape, a "clean" and properly worked landscape, an agricultural landscape?

11.5. A confusion between education forms in the ABR: formal, non-formal and informal

Three education forms (formal, non-formal and informal) enable us to question learning, knowledge, places and environments in the case of the ABR around Taroudant in Morocco.

Let us recall that the formal education generally offered at the biosphere reserve (BR) takes place in educational and training institutions and leads to the acquisition of recognized diplomas and qualifications.

From a higher education viewpoint in Morocco, university courses on the ABR are thematic; they do not correspond to a discipline in the strict sense, but rather constitute modular education on patrimony, the environment and sustainable development. The ABR-related formal education offered in Moroccan universities has four characteristics, as observed by the authors (Barthes et al. 2017):

– it is linked to socially acute questions debated in society;

– it gives an important place to values;

– it aims to change the behavior, goals, values and knowledge at play beyond the school framework;

– it combines theoretical and practical lessons (textual objects, pictures, educational films, scientific publications, etc.).

To this formal education at the ABR is opposed the "non-formal education" taking place outside the main education and training structures, which does not usually lead to the acquisition of official certificates. Non-formal education can be acquired at the workplace or through the activities of organizations or civil society groups (youth associations, NGOs, etc.).

Today, non-formal education offered at the ABR and its mediation calls upon several types of "local knowledge" from "naturalists" or "peasants":

– knowledge related to plants (cultivated, wild, food, medicinal, etc.);

– forest-related knowledge;

– knowledge on animal husbandry, culture;

– crafts knowledge.

It also draws on local knowledge related to (technical and ecological) management and its uses and transformations:

– organizational knowledge (*agdals*[1] or "*AOC*" designations);

– relational knowledge (knowing how to regulate the relationships of individuals or groups in relation to the access to and the use of resources);

– knowledge "on the outside" (the perception of non-local actors).

1 The *agdal* is a way of community management characterized by the defense of resources within a delimited territory (Auclair et al. 2010).

All of this ABR local knowledge or non-formal education is inseparable from the representations of the natural environment, as well as from religious, cultural, linguistic factors, etc. It is both produced by a collective over the long term and modified by each bearer, transmitted "in situ" in a non-formal way, often also in an intergenerational manner. It can draw upon "common" local knowledge from a set of agents (a tribe, a group of "professionals": therapists, beekeepers, etc.).

The non-formal education offered at the ABR, like the local knowledge, is part of the currents of interpretation and education based on the feelings of the apprentices or their experiences and the appropriation of the environment. It is opposed to the way in which school knowledge and learning take place inside universities with a prescribed curriculum based on rationality rather than affect. A partnership could be established between the university and the patrimonial ABR, between those responsible for education and tourism, through better research on scholarly knowledge and the knowledge to be taught.

A first observation on the ABR: there is a hybridization between formal and non-formal education (Barthes and Alpe 2018), between the school form (the prescribed curriculum) and the school culture (all the knowledge produced for and by the school) (Chervel 1998).

A second observation: there is a dissemination of knowledge from the extra-curricular environment towards the school environment and the other way round. Insofar as non-formal knowledge is often less recognized by the school institution, the patrimonial ABR seeks to fit this knowledge into the curricula, recognizing it has passed the test of legitimacy (Alpe 2006).

Another category is added to these forms: informal education, the natural corollary of daily life. Unlike formal and non-formal education, it is not necessarily intentional and may not be recognized, even by individuals themselves, as an addition to their knowledge and their skills.

From the point of view of education, and critical reflection on the content of digital mediation devices, from the perspective questioning the didactic transposition of scholarly knowledge to the knowledge to be taught, it appears that experts-teachers involved in education at the ABR rely just as much on their position as teachers (the prescribed curriculum) as on their knowledge as researchers (formal, non-formal or informal). This transmitted expertise can be implicitly or explicitly experienced by the students as a barrier to their own learning and recreation. University expertise on the material to be taught – in particular virtual ABR education – as well as the posture adopted by the teachers-researchers are fundamentally placed at the service of a digital mechanism focusing on professional development.

It is necessary to properly frame this expertise as a strategic asset in a broader system, acknowledging the profession's constraints and the legitimation of knowledge (or not). Conversely, educational mechanisms can be perceived as spaces for engaging the students' interest in human sciences knowledge: learning a research and questioning approach takes on further meaning, insofar as it becomes significant for the students because they re-problematize knowledge, the objects of their creation or their purpose. Ultimately, the purpose is to engage the interest of students in patrimony-based education at the ABR. The question does not merely involve digitizing learning content, but rather creating a movement of reflection, of uses, of creation around this new "digital patrimony" under construction and of considering a real mediation action engaging students in their learning process in practical terms.

For the ABR virtual environment to become properly educational, it will first be necessary to invent educational technologies through:

– a cultural revolution:

- the virtual endows the production of knowledge with the risk of drifting away from reality;

- definitively abandon the illusion that technologies will teach and train on their own;

- if everything in nature is a lesson, we must re-examine knowledge and practice openness to the lessons that the world can transmit to us; the territory thus becomes a source of innovative lessons;

- exploit the impact of IT tools on patrimony-based education;

– research and training:

- develop theoretical research in the education sciences promoting the advent of efficient educational technologies;

- manage the design and pedagogic communication of educational IT environments;

– technological development and economic investment:

- develop hardware and software interfaces that can lie low;

- take the political decision to equip all schools, especially the most disadvantaged ones, with microcomputers connected to networks;

- grasp the different forms of education (formal, non-formal and informal) and adopt new technologies to pedagogical teaching practices.

Finally, for researchers, we can adopt two forms of education (formal and non-formal) inspired by different educational institutions, such as the school and the museum.

School	Museum
– Education based on a prescribed curriculum: - programming - progression of learning - legitimized knowledge - texts, tools and media – Goals focusing on skills: - knowledge acquisition - capacities acquisition - attitudes acquisition – Subject: the student, the evaluated individual *Cognition register*	– Education often based on exposure: - narrative framework - background: impact of learning - knowledge chosen depending on a point of view - texts, objects, space – Goals dependent on biases: - to teach - to debate - to make feel – Subject: the visitor, appreciative individual *Register of delight*

Table 11.1. *University and museum: two different educational institutions*

11.6. Towards mediation in the ABR or the construction of an alliance and communication strategies between education and tourism actors

The actor approach was chosen to carry out the empirical phase with an exploratory and comprehensive purpose. An applied research posture was adopted. It was centered on three actors responsible for mediation of the ABR:

– UNESCO, as an international organization, carries out actions favoring communication and alliance strategies with other actors with a view to fostering education, mediation and media coverage of the ABR in Morocco. It funds sustainable development projects and contributes to the creation of professional networks which make these orientations visible.

– Tourism actors prompt the valorization and socio-economic development of the ABR through mediation.

– Education actors develop skills in terms of learning content development and digital mediation of the ABR.

This illustrates the high potential for political support that these three actors can gain from an alliance for establishing committed communication favoring education, mediation and media coverage of the ABR.

These actors can be located under another typology related to communication, which characterizes the central actors who carry out both educational actions and digital mediation of the ABR in their missions or by strategic choice. The alliance between actors can also be shown by a so-called basic strategic diagram, describing a game between four types of actors:

– the two allies;

– a “target actor” (A-c) whose allies want to influence behavior;

– a “third-party actor” (A-t), whose function is not specified a priori. We assume that the third-party actor can be either a regulator, a prescriber or a mixed actor, a tourism actor (TA) or an education actor (EA), including UNESCO.

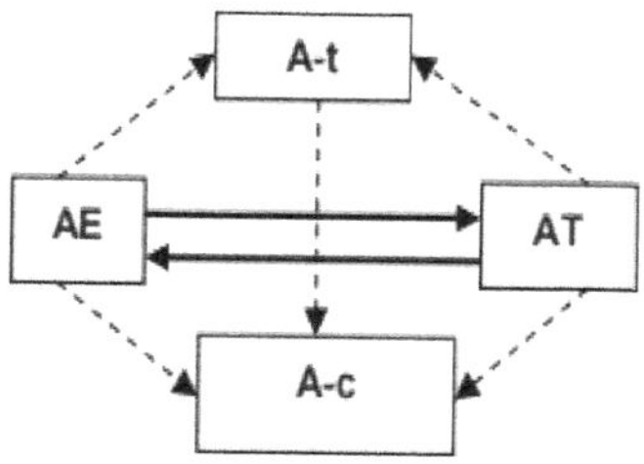

Figure 11.1. *Actor modeling*

The logics of the actors taking part in ABR mediation are analyzed in the sphere of intentional activities (we must not forget the importance of actors who have unintentional impact in one field or another).

All of these ABR mediators perceive their responsibility in intentional management as being an educational activity which rejoins the purpose of the tourism sector to reach the 2030 Sustainable Development Goals.

We analyze this particular profile through the prism of organizations management, in particular M. Marchesnay’s grid (1993), which articulates the examination of the nature of an organization’s activities (transversality and innovation), the organization’s structure, its multidisciplinary teams and their relationships with their environments.

Analysis of the intentional management of actors operating the ABR educational and digital mediation		
Intentional management of mediation by missions	UNESCO	Researchers
Intentional management of mediation via strategy	Actors in education, BRs, tourism	Civil society
Institutional actors in education Institutional actors in tourism	Intentional activity by strategy	Intentional activity by mission

Table 11.2. *Analysis of the intentional management of actors operating the ABR education and digital mediation*

We can draw up a typology of educational activities in the field of biosphere reserves mediation through an analysis of actors' games. The activities are the following:

– "niche" activities exploiting biosphere reserves in the field of "ecotourism";

– alliance and communication activities;

– educational, mediation and media coverage activities;

– education proactive activities and strategies to be implemented for human and sustainable development.

As regards the trade-offs between partially antagonistic goals, UNESCO could sit at the center of a partners network structure to strike a balance between different fields, especially education and tourism.

In broader terms, this vision of the actor network is the most consistent with the exercise of animation and demonstration missions by UNESCO as part of the skills development project for education and applied research on Mediterranean biosphere reserves (2018–2021).

11.7. The territorial integration of the ABR – a condition for the alliance's success: communication, mediation and media coverage

Territorial integration requires an alliance and committed communication between the actors responsible for education, tourism, the ABR and the local population for the territorial management of the ABR and its mediation. It can only be strengthened by being part of a perspective concerning human and sustainable development. The style and form of communication between the actors should favor the ABR's territorial integration. These are essential elements of the strategy sought for education, mediation and media coverage in the ABR.

We rely on the concept of the "service triangle". It distinguishes between three poles: the customer, the service provider and a "support-good" behind the service, in this case, the ABR. The service provider includes the largest number of "actors" in the territory and beyond, intentionally or unintentionally taking part in ABR education and mediation. We extend this model to acknowledge the prescribing role of the customer, the user and, above all, the roles of the population in relation to patrimony. Our model incorporates the notion of the "ABR controller" in the sense of a synergy of relationships between the actors' education and tourism. Thus, the "actors" pole is the most complete, and its modeling crosses the functional and intentional dimensions, enabling strategic reasoning.

The territorial servuction[2] model is schematized in Figure 11.2.

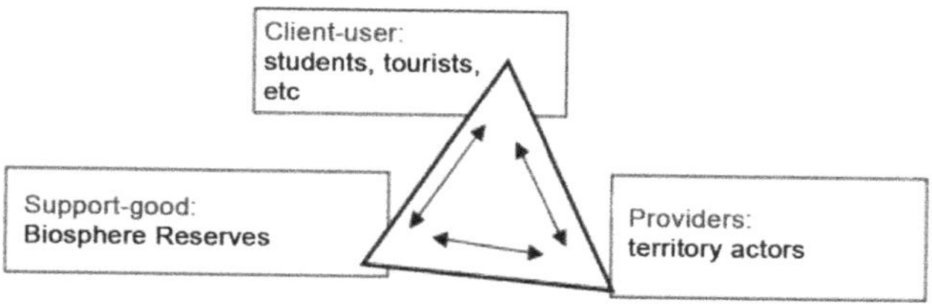

Figure 11.2. *Territorial servuction model*

The complete model of the "provider" pole leads to a second triangle, similar to the previous one, situating the main actor functions in relation to the three initial poles: the client-user (students, tourists), the support-good (the ABR) and the service provider (territorial actors). This second triangle shifts from the previous one by symbolizing the fact that the actors in the center of the superposition (dotted area), in particular the actors responsible for education, the ABR, tourism and the international organization UNESCO, perform their function intentionally, while the others, namely the population, do so unintentionally.

This proposed model leads to functional (the service triangle, the types of actors defined by their roles), evaluative (goal dimensions) and pre-strategic reflection (through the distinction relating to the intentionality of the roles).

2 The concept of servuction was developed by Pierre Eiglier and Éric Langeard. It represents all the material and human elements used, as well as the activities deployed, to design, create and develop the service a company wishes to offer on the market, in accordance with a chosen quality level.

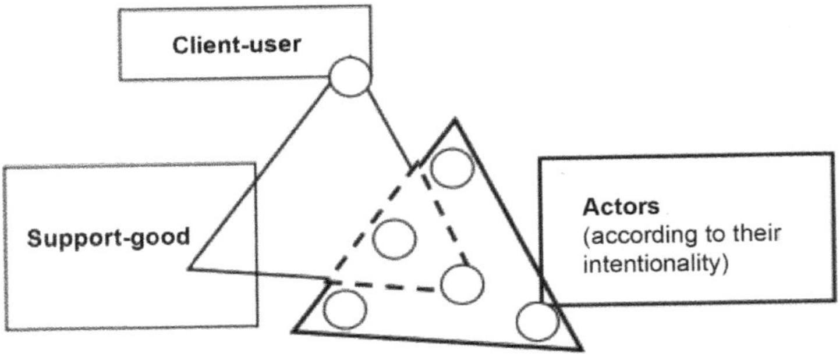

Figure 11.3. *Functional model*

11.8. "Polarized" networks in the ABR: a tool for the alliance between education and tourism actors

To characterize the functions that the ABR network tool can bring to the mediation and alliance of education and tourism actors, we can note four methods that can be used for setting up bridges of communication between the actors:

– the "communication" established between UNESCO and education and tourism actors;

– "cognition" by modifying the representations and influencing the logic of the actors;

– the (joint or concerted) "realization", specifically the "implementation", of the choice of educational forms to be communicated and mediated;

– "political support" for the territorial integration of the ABR: supporting a proactive vision for the development of skills in ABR education, mediation and media coverage in the long term.

The distinction between two relational configurations (three structures of the global network connecting ABR actors) could contribute to managing communication and mediation via a planned information system:

– the UNESCO and education actors building up a formal network (e.g. a communication group), called "core network";

– tourism and environmental actors in contact with the core network as a whole;

– all the actors forming a single network.

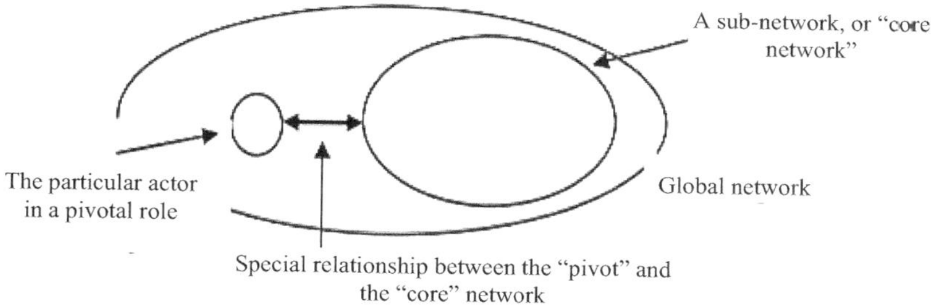

Figure 11.4. *Networks and actors*

As regards the ABR actor games, the logics of their participation in the model could be taken into account according to three dimensions:

– the utilitarian interest of collective action;

– the commitment to human values;

– the attraction of social relations.

The strategies of ABR actors should acknowledge the fact that:

– networking is a tool for forming alliances between actors in a win–win manner;

– UNESCO's strategy is useful for launching development and preservation processes;

– UNESCO can act on critical factors (trust, consensus on the shared project, rules, relationship costs, performance, etc.) and oversee the strategic renewal of the network actors;

– the State is an architect and coordinator of various actions in the ABR. It is increasingly territorializing its actions by encouraging the commitment of local actors.

11.9. Actor training for the development of capacities: skills and capability for communication management

In this context, capacity designates the quality of an individual to understand and to do something: being able to do something as long as one has the qualities and the resources to do so. Capacity is related to the act, the possibility and the force or energy for doing something; it refers to the ability to perform a certain action (physical, behavioral, mental, etc.). In contrast, the competence or skill is related to in-depth knowledge (about a subject, discipline, etc.); it designates the qualities required for the performance of a task, of a function. It is the individual's cognition regarding knowledge and mental faculties. Strengthening the interaction between these three "C"s (competence or skill, capacity, capability) through training favors the exploitation of the actors' know-how in communication with a certain flexible degree of freedom of choice in the action, relying on their capacities and competences or skills.

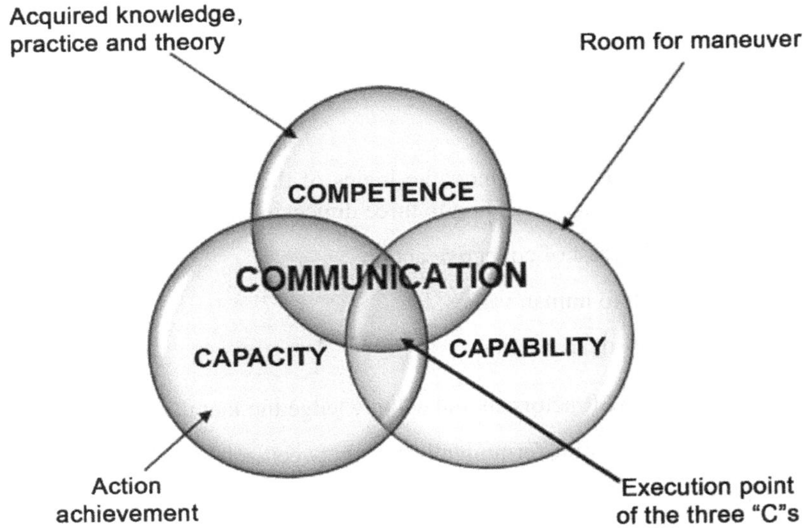

Figure 11.5. *Reinforcement of the interactions between the three "C"s (competence or skill, capacity and capability) through training and communication management between actors (Source: authors' elaboration 2021)*

Top-down and bottom-up communication and its effectiveness depend on the importance of the dimension of the zone of interaction between the three "C"s among multilevel actors exerting their individual and collective capabilities. The emphasis is also placed on the relationship between the actor and power. Power always has its own favorable position in any collective activity. It is fundamental

> because power constitutes a daily mechanism in our social existence. We constantly use it in our relations with our friends, our colleagues, our family, etc. [...]. Suppressing power means nothing other, in the end, than suppressing the possibility for but also the right of actors to do something other than what is expected of them, in short, suppressing their autonomy to reduce them to a machine condition (Crozier and Freidberg 1992, p. 33).

11.10. Conclusion

This study has a marked exploratory character, and this within the broad complex field of educational forms offered at the ABR in Morocco: it nourishes knowledge on the ABR in the field of territorial management, and in particular, it highlights the role of alliance and communication between actors in education and mediation. It proposes an approach for the construction of alliance strategies applicable to education and tourism actors. This approach is both systemic and dialogical in that it articulates "private" alliance strategies and globally concerted management strategies. This patrimony-based education model questions the construction of a strategic theory of polycentric collective action.

The reflection pursued in this work has led us to analyze the integrated approach of UNESCO's biosphere reserves via its world heritage sites. According to this approach, communication is an instrument for encouraging development, education, environmental protection, tourism sustainability, etc. While the approach chosen was inspired by the results of studies accepting the idea of patrimony mediation as something done ex post, in our theoretical reflections on human-based and sustainable development, we consider it as something that should be pursued ex ante.

In patrimony-based digital education, the student sensitive to the limits of their own perception systems becomes attentive to virtual realities. The same process also enables the indigenous inhabitant to take a distanced look at their own patrimonial wealth, defined on the basis of their own reality, rather than with reference to other types of wealth (likely to be found elsewhere). Even though we live in one unique connected world, this world is nonetheless unique only in virtue of its patrimonial diversity. Education needs to embrace such diversity, and possibly accept that there may be different contents or knowledge for each container. At the same time, it is essential to use digital instruments adapted to the reality of the ABR's territorial integration in Morocco. Such a configuration can only be envisaged by daring to "open up ourselves, while inserting this openness into a wider openness to the world where we live" (Eberhard 2014).

Finally, we are aware that it is necessary to shift from "informed communication" to "committed communication". This is key for the strategic transformation of education and for improving mediation and media coverage of the ABR.

11.11. References

Alpe, Y. (2006). Quelle est la légitimité des savoirs scolaires ? In *L'école à l'épreuve de l'actualité : enseigner les questions vives*, Legardez, A. and Simonneaux, L. (eds). ESF, Issy-les-Moulineaux.

Auclair, L., Simenel, R., Alifriqui, M., Michon, G. (2010). "Agdal" les voies Imazighen de la patrimonialisationdu territoire. *Hesperis-Tamuda*, 45, 129–149.

Barouch, G. (1989). *La décision en miettes*. L'Harmattan, Paris.

Barthes, A. (2017). Éducation au patrimoine. In *Introduction. Dictionnaire critique des enjeux et concepts des "éducations à"*, Barthes, A., Lange, J.M., Tutiaux-Guillon, N. (eds). L'Harmattan, Paris.

Barthes, A. and Alpe, Y. (2015). L'éducation au patrimoine dans les aires territoriales protégées, une dimension de l'éducation au développement durable ? *Éducation au développement durable*, Diemer, A. and Marquat, C. (eds). De Boeck Supérieur, Brussels.

Barthes, A. and Alpe, Y. (2018). Les "éducations à", une remise en cause de la forme scolaire ? *Carrefours de l'éducation*, 45(1), 23–37.

Barthes, A., Lange, J.M., Tutiaux-Guillon, N. (eds) (2017). *Dictionnaire critique des enjeux et concepts des "éducations à"*. L'Harmattan, Paris.

Chervel, A. (1998). *La culture scolaire : une approche historique*. Belin, Paris.

Crozier, M. and Friedberg, E. (1992). *L'acteur et le système : les contraintes de l'action collective*. Le Seuil, Paris.

David, A. (2004). Étude de cas et généralisation scientifique en sciences de gestion. *13e conférence de l'AIMS*, AIMS, Normandie-vallée de Seine.

Dubost, F. and Lizet, B. (2003). *La nature dans la cité, de l'hygiénisme au développement durable*. Persée, Lyon.

Eberhard, C. (2014). *Oser le plurivers : pour une globalisation interculturelle et responsable*. Connaissances et Savoirs, Paris.

Friedberg, E. (1993). *Le pouvoir et la règle – Dynamiques de l'action organisée*. Le Seuil, Paris.

Kadri, B. (2012). Réflexions sur l'épistémologie du tourisme : la perspective constructiviste. In *Épistémologie des études touristiques*, Morisset, L.K., Sarrasin, B., Éthier, G. (eds). PUQ, Quebec.

Leininger-Frézal, C. (2009). Le développement durable et ses enjeux éducatifs. Acteurs, savoirs et stratégies territoriales. PhD thesis, Lumière University Lyon 2, Lyon.

Levêque, C. (2001). *Écologie. De l'écosystème à la biosphère*. Dunod, Paris.

Marchesnay, M. (1993). *Management stratégique*. Eyrolles, Paris.

Ollagnon, H. (2001). Stratégies patrimoniales pour un développement durable. PhD thesis, University of Reims Champagne-Ardenne, Reims.

Romagny, B. and Boujrouf, S. (2010). La ruée vers l'huile d'argan, chronique de la patrimonialisation d'un terroir marocain. *Hespéris Tamuda*, 45, 151–166.

List of Authors

Hannah ABOU FAKHER
International Federation of Red Cross and Red Crescent Societies
Beirut
Lebanon

Mohammed ADERGHAL
Mohammed V University
Rabat
Morocco

Abdullah AÏT L'HOUSSAIN
Cadi-Ayyad University
Marrakech
Morocco

Lahoucine AMZIL
Mohammed V University
Rabat
Morocco

Nina ASLOUM
ENSFEA
Toulouse
France

Didier BABIN
MAB Programme
UNESCO
Paris
France

Angela BARTHES
University of Aix-Marseille
Digne
France

Laurent BEDOUSSAC
ENSFEA
Toulouse
France

Sylviane BLANC-MAXIMIN
University of Aix-Marseille
Digne
France

Éliane BOU DAGHER
Saint-Joseph University
Beirut
Lebanon

Meriem BOUAMRANE
Head of Biodiversity and Ecology
MAB Programme
UNESCO
Paris
France

Magda BOU DAGHER KHARRAT
Saint-Joseph University
Beirut
Lebanon

Saïd BOUJROUF
Cadi-Ayyad University
Marrakech
Morocco

Véronique CHALANDO
University of Aix-Marseille
Digne
France

Mikaël CHAMBRU
Grenoble Alpes University
France

Catherine CIBIEN
MAB Programme
UNESCO
Paris
France

Cécilia CLAEYS
University of Perpignan
France

Wassim EL-HAJJ
American University of Beirut
Lebanon

Bruno GARNIER
CNRS
University of Corsica
Corsica

Guillaume GILLET
ENSFEA
Toulouse
France

Moustapha ITANI
American University of Beirut
Lebanon

Salma ITSMAÏL
CNRS
University of Corsica
Corsica

Rhéa KAHALÉ
Saint-Joseph University
Beirut
Lebanon

Jean-Marc LANGE
University of Montpellier
France

Roser MANEJA
Autonomous University of Barcelona
Spain

Salma NASHABE TALHOUK
American University of Beirut
Lebanon

Nivine NASRALLAH
American University of Beirut
Lebanon

Bruno ROMAGNY
Institut de recherche pour le développement (IRD)
Marseille
France

Melki SLIMANI
ISEAH
Sbeïtla
University of Kairouan
Tunisia

Index

G, H, I

K, L

M, N

P, R

S, T

U, Z

Summary of Volume 2

Chapter 5. Do We Need to Choose Between Biodiversity, Industry and Tourism? A Metabolic Approach to Manage the Mediterranean Biosphere Reserve of Menorca
Alejandro MARCOS-VALLS

Chapter 6. The Jabal Moussa Biosphere Reserve (Lebanon): A Private Association Initiative
Pierre DOUMET and Joelle BARAKAT

Chapter 7. Understandings of Administration and Challenges to Governance in the Arganeraie Biosphere Reserve (Morocco)
Abdelaziz AFKER

Chapter 16. The Legacy and Future of Conservation in El Kala National Park (Algeria)
Tarik DAHOU

Chapter 17. Social Representations of Biospheres and Sustainable Local Development in Bou Hedma (Tunisia)
Abdelkarim BRAHMI

Chapter 18. Architecture and the Biosphere Environment in Pedagogy: Design Visions for Sustainable Dwelling Communities

Carla ARAMOUNY

Other titles from

iSTE

in

Science, Society and New Technologies

2023

CARDON Alain, EL HAMI Abdelkhalak
Fundamental Generation Systems: Computer Science and Artificial Consciousness, the Informational Field of Generation of the Universe, the Sixth Sense of Living Beings
(Digital Sciences Set – Volume 4)

ELAMÉ Esoh
The Sustainable City in Africa Facing the Challenge of Liquid Sanitation
(Territory Development Set – Volume 2)

JURCZENKO Emmanuel
Climate Investing: New Strategies and Implementation Challenges

PARET Dominique, CRÉGO Pierre, SOLÈRE Pauline
Smart Patches: Biosensors, Graphene and Intra-Body Communications

VENTRE Daniel, LOISEAU Hugo
Cybercrime during the SARS-CoV-2 Pandemic (2019-2022): Evolutions, Adaptations, Consequences
(Cybersecurity Set – Volume 3)

2022

AIT HADDOU Hassan, TOUBANOS Dimitri, VILLIEN Philippe
Ecological Transition in Education and Research

CARDON Alain
Information Organization of The Universe and Living Things: Generation of Space, Quantum and Molecular Elements, Coactive Generation of Living Organisms and Multiagent Model
(Digital Science Set – Volume 3)

CAULI Marie, FAVIER Laurence, JEANNAS Jean-Yves
Digital Dictionary

DAVERNE-BAILLY Carole, WITTORSKI Richard
Research Methodology in Education and Training: Postures, Practices and Forms
(Education Set – Volume 12)

ELAMÉ Esoh
Sustainable Intercultural Urbanism at the Service of the African City of Tomorrow
(Territory Development Set – Volume 1)

FLEURET Sébastien
A Back and Forth Between Tourism and Health: From Medical Tourism to Global Health
(Tourism and Mobility Systems Set – Volume 5)

KAMPELIS Nikos, KOLOKOTSA Denia
Smart Zero-energy Buildings and Communities for Smart Grids
(Engineering, Energy and Architecture Set – Volume 9)

2021

BARDIOT Clarisse
Performing Arts and Digital Humanities: From Traces to Data
(Traces Set – Volume 5)

BENSRHAIR Abdelaziz, BAPIN Thierry
From AI to Autonomous and Connected Vehicles: Advanced Driver-Assistance Systems (ADAS)
(Digital Science Set – Volume 2)

BÉRANGER Jérôme
Societal Responsibility of Artificial Intelligence: Towards an Ethical and Eco-responsible AI
Technological Prospects and Social Applications Set – Volume 4)

CORDELIER Benoit, GALIBERT Oliver
Digital Health Communications
Technological Prospective and Social Applications Set – Volume 5)

DOUAY Nicolas, MINJA Michael
Urban Planning for Transitions

GALINON-MÉLÉNEC Béatrice
The Trace Odyssey 1: A Journey Beyond Appearances
(Traces Set – Volume 4)

HENRY Antoine
Platform and Collective Intelligence: Digital Ecosystem of Organizations

KOLOKOTSA Denia, KAMPELIS Nikos
Smart Buildings, Smart Communities and Demand Response
(Engineering, Energy and Architecture Set – Volume 8)

LE LAY Stéphane, SAVIGNAC Emmanuelle, LÉNEL Pierre, FRANCES Jean
The Gamification of Society
(Research, Innovative Theories and Methods in SSH Set – Volume 2)

RADI Bouchaïb, EL HAMI Abdelkhalak
Optimizations and Programming: Linear, Non-linear, Dynamic, Stochastic and Applications with Matlab
(Digital Science Set – Volume 1)

2020

ALAKTIF Jamila, CALLENS Stéphane
Migration and Climate Change: From the Emergence of Human Cultures to Contemporary Management in Organizations

BARNOUIN Jacques
The World's Construction Mechanism: Trajectories, Imbalances and the Future of Societies
(Interdisciplinarity between Biological Sciences and Social Sciences Set – Volume 4)

ÇAĞLAR Nur, CURULLI Irene G., SIPAHIOĞLU Işıl Ruhi, MAVROMATIDIS Lazaros
Thresholds in Architectural Education
(Engineering, Energy and Architecture Set – Volume 7)

DUBOIS Michel J.F.
Humans in the Making: In the Beginning was Technique
(Social Interdisciplinarity Set – Volume 4)

ETCHEVERRIA Olivier
The Restaurant, A Geographical Approach: From Invention to Gourmet Tourist Destinations
(Tourism and Mobility Systems Set – Volume 3)

GREFE Gwenaëlle, PEYRAT-GUILLARD Dominique
Shapes of Tourism Employment: HRM in the Worlds of Hotels and Air Transport
(Tourism and Mobility Systems Set – Volume 4)

JEANNERET Yves
The Trace Factory
(Traces Set – Volume 3)

KATSAFADOS Petros, MAVROMATIDIS Elias, SPYROU Christos
Numerical Weather Prediction and Data Assimilation
(Engineering, Energy and Architecture Set – Volume 6)

MARTI Caroline
Cultural Mediations of Brands: Unadvertization and Quest for Authority
(Communication Approaches to Commercial Mediation Set – Volume 1)

MAVROMATIDIS Lazaros E.
Climatic Heterotopias as Spaces of Inclusion: Sew Up the Urban Fabric
(Research in Architectural Education Set – Volume 1)

MOURATIDOU Eleni
Re-presentation Policies of the Fashion Industry: Discourse, Apparatus and Power
(Communication Approaches to Commercial Mediation Set – Volume 2)

SCHMITT Daniel, THÉBAULT Marine, BURCZYKOWSKI Ludovic
Image Beyond the Screen: Projection Mapping

VIOLIER Philippe, with the collaboration of TAUNAY Benjamin
The Tourist Places of the World
(Tourism and Mobility Systems Set – Volume 2)

2019

BRIANÇON Muriel
The Meaning of Otherness in Education: Stakes, Forms, Process, Thoughts and Transfers
(Education Set – Volume 3)

DESCHAMPS Jacqueline
Mediation: A Concept for Information and Communication Sciences
(Concepts to Conceive 21st Century Society Set – Volume 1)

DOUSSET Laurent, PARK Sejin, GUILLE-ESCURET Georges
Kinship, Ecology and History: Renewal of Conjunctures
(Interdisciplinarity between Biological Sciences and Social Sciences Set – Volume 3)

DUPONT Olivier
Power
(Concepts to Conceive 21st Century Society Set – Volume 2)

FERRARATO Coline
Prospective Philosophy of Software: A Simondonian Study

GUAAYBESS Tourya
The Media in Arab Countries: From Development Theories to Cooperation Policies

HAGÈGE Héléne
Education for Responsibility
(Education Set – Volume 4)

LARDELLIER Pascal
The Ritual Institution of Society
(Traces Set – Volume 2)

LARROCHE Valérie
The Dispositif
(Concepts to Conceive 21st Century Society Set – Volume 3)

LATERRASSE Jean
Transport and Town Planning: The City in Search of Sustainable Development

LELEU-MERVIEL Sylvie, SCHMITT Daniel, USEILLE Philippe
From UXD to LivXD: Living eXperience Design

LENOIR Virgil Cristian
Ethically Structured Processes
(Innovation and Responsibility Set – Volume 4)

LOPEZ Fanny, PELLEGRINO Margot, COUTARD Olivier
Local Energy Autonomy: Spaces, Scales, Politics
(Urban Engineering Set – Volume 1)

METZGER Jean-Paul
Discourse: A Concept for Information and Communication Sciences
(Concepts to Conceive 21st Century Society Set – Volume 4)

MICHA Irini, VAIOU Dina
Alternative Takes to the City
(Engineering, Energy and Architecture Set – Volume 5)

PÉLISSIER Chrysta
Learner Support in Online Learning Environments

PIETTE Albert
Theoretical Anthropology or How to Observe a Human Being
(Research, Innovative Theories and Methods in SSH Set – Volume 1)

PIRIOU Jérôme
The Tourist Region: A Co-Construction of Tourism Stakeholders
(Tourism and Mobility Systems Set – Volume 1)

PUMAIN Denise
Geographical Modeling: Cities and Territories
(Modeling Methodologies in Social Sciences Set – Volume 2)

WALDECK Roger
Methods and Interdisciplinarity
(Modeling Methodologies in Social Sciences Set – Volume 1)

2018

BARTHES Angela, CHAMPOLLION Pierre, ALPE Yves
Evolutions of the Complex Relationship Between Education and Territories
(Education Set – Volume 1)

BÉRANGER Jérôme
The Algorithmic Code of Ethics: Ethics at the Bedside of the Digital Revolution
(Technological Prospects and Social Applications Set – Volume 2)

DUGUÉ Bernard
Time, Emergences and Communications
(Engineering, Energy and Architecture Set – Volume 4)

GEORGANTOPOULOU Christina G., GEORGANTOPOULOS George A.
Fluid Mechanics in Channel, Pipe and Aerodynamic Design Geometries 1
(Engineering, Energy and Architecture Set – Volume 2)

GEORGANTOPOULOU Christina G., GEORGANTOPOULOS George A.
Fluid Mechanics in Channel, Pipe and Aerodynamic Design Geometries 2
(Engineering, Energy and Architecture Set – Volume 3)

GUILLE-ESCURET Georges
Social Structures and Natural Systems: Is a Scientific Assemblage Workable?
(Social Interdisciplinarity Set – Volume 2)

LARINI Michel, BARTHES Angela
Quantitative and Statistical Data in Education: From Data Collection to Data Processing
(Education Set – Volume 2)

LELEU-MERVIEL Sylvie
Informational Tracking
(Traces Set – Volume 1)

SALGUES Bruno
Society 5.0: Industry of the Future, Technologies, Methods and Tools
(Technological Prospects and Social Applications Set – Volume 1)

TRESTINI Marc
Modeling of Next Generation Digital Learning Environments: Complex Systems Theory

2017

ANICHINI Giulia, CARRARO Flavia, GESLIN Philippe,
GUILLE-ESCURET Georges
Technicity vs Scientificity – Complementarities and Rivalries
(Interdisciplinarity between Biological Sciences and Social Sciences Set – Volume 2)

DUGUÉ Bernard
Information and the World Stage – From Philosophy to Science, the World of Forms and Communications
(Engineering, Energy and Architecture Set – Volume 1)

GESLIN Philippe
Inside Anthropotechnology – User and Culture Centered Experience
(Social Interdisciplinarity Set – Volume 1)

GORIA Stéphane
Methods and Tools for Creative Competitive Intelligence

KEMBELLEC Gérald, BROUDOUS Evelyne
Reading and Writing Knowledge in Scientific Communities: Digital Humanities and Knowledge Construction

MAESSCHALCK Marc
Reflexive Governance for Research and Innovative Knowledge (Responsible Research and Innovation Set - Volume 6)

PARK Sejin, GUILLE-ESCURET Georges
Sociobiology vs Socioecology: Consequences of an Unraveling Debate (Interdisciplinarity between Biological Sciences and Social Sciences Set – Volume 1)

PELLÉ Sophie
Business, Innovation and Responsibility (Responsible Research and Innovation Set – Volume 7)

2016

ANDRÉ Michel, SAMARAS Zissis
Energy and Environment (Research for Innovative Transports Set – Volume 1)

BLANQUART Corinne, CLAUSEN Uwe, JACOB Bernard
Towards Innovative Freight and Logistics (Research for Innovative Transports Set – Volume 2)

BRONNER Gérald
Belief and Misbelief Asymmetry on the Internet

COHEN Simon, YANNIS George
Traffic Management (Research for Innovative Transports Set – Volume 3)

EL FALLAH SEGHROUCHNI Amal, ISHIKAWA Fuyuki, HÉRAULT Laurent, TOKUDA Hideyuki
Enablers for Smart Cities

GIANNI Robert
Responsibility and Freedom (Responsible Research and Innovation Set – Volume 2)

GRUNWALD Armin
The Hermeneutic Side of Responsible Research and Innovation
(Responsible Research and Innovation Set – Volume 5)

LAGRAÑA Fernando
E-mail and Behavioral Changes: Uses and Misuses of Electronic Communications

LENOIR Virgil Cristian
Ethical Efficiency: Responsibility and Contingency
(Responsible Research and Innovation Set – Volume 1)

MAESSCHALCK Marc
Reflexive Governance for Research and Innovative Knowledge
(Responsible Research and Innovation Set – Volume 6)

PELLÉ Sophie, REBER Bernard
From Ethical Review to Responsible Research and Innovation
(Responsible Research and Innovation Set – Volume 3)

REBER Bernard
Precautionary Principle, Pluralism and Deliberation: Sciences and Ethics
(Responsible Research and Innovation Set – Volume 4)

TORRENTI Jean-Michel, LA TORRE Francesca
Materials and Infrastructures 1
(Research for Innovative Transports Set – Volume 5A

TORRENTI Jean-Michel, LA TORRE Francesca
Materials and Infrastructures 2
(Research for Innovative Transports Set – Volume 5B)

VENTRE Daniel
Information Warfare – 2nd edition

YANNIS George, COHEN Simon
Traffic Safety
(Research for Innovative Transports Set – Volume 4)

Printed and bound by CPI Group (UK) Ltd, Croydon, CR0 4YY
16/09/2026
14963088-0004